Machine Learning and Systems Biology in Genomics and Health

Shailza Singh
Editor

Machine Learning and Systems Biology in Genomics and Health

Springer

Editor
Shailza Singh
Department of Pathogenesis and Cellular
Response
National Centre for Cell Science
Pune, Maharashtra, India

ISBN 978-981-16-5992-8 ISBN 978-981-16-5993-5 (eBook)
https://doi.org/10.1007/978-981-16-5993-5

This Springer imprint is published by the registered company Springer Nature Singapore Pte Ltd.
The registered company address is: 152 Beach Road, #21-01/04 Gateway East, Singapore 189721, Singapore

Contents

About the Editor

Shailza Singh is Scientist-E and Incharge of Bioinformatics and High Performance Computing Facility, National Centre for Cell Science, Pune, India. Her research chiefly focuses on systems and synthetic biology. She also specializes in infectious diseases such as leishmaniasis. Her research group is working to integrate the action of regulatory circuits, cross-talk between pathways, and nonlinear kinetics of biochemical processes through mathematical modeling. Dr. Singh has been honored with the DBT-RGYI, DST Young Scientist, and INSA Bilateral Exchange Programme awards and was selected by the DBT for a SAKURA EXCHANGE Programme in Science in the field of Artificial Intelligence and Machine Learning to Tokyo in 2018. She serves as a reviewer for prestigious international grants such as the Research Councils UK; for national grants from the DBT, DST, and CSIR; and for several prominent international journals.

Construction of Feedforward Multilayer Perceptron Model for Diagnosing Leishmaniasis Using Transcriptome Datasets and Cognitive Computing

1

M. A. Sundaramahalingam, Ritika Kabra, and Shailza Singh

Abstract

Leishmaniasis is an endemic parasitic disease, predominantly found in the poor locality of Africa, Asia, and Latin America. It is associated with malnutrition, the weak immune system of people, and their housing locality. It is diagnosed by microscopic identification, molecular and biochemical characterization, or serum analysis for parasitic compounds. In this study, we present a new approach for diagnosing leishmaniasis using transcriptome data and cognitive computing. The transcriptome data of leishmaniasis were collected from the Gene Expression Omnibus database, and it was processed. The algorithm for training and developing a model based on the data was prepared and coded using python. The algorithm and their corresponding datasets were integrated using the TensorFlow data frame. A feedforward artificial neural network trained model with multilayer perceptron was developed as a diagnosing model for leishmaniasis, using transcriptome data. It was created using a recurrent neural network. The cognitive model of the trained network was interpreted using the maps and mathematical formula of the influencing parameters. The credit of the system was measured using the accuracy, loss, and error of the system. This integrated leishmaniasis transcriptome data and neural network system acted as an excellent diagnosis model with higher accuracy and lower error. The experimental results of

M. A. Sundaramahalingam
Department of Biotechnology, Kamaraj College of Engineering and Technology, Madurai, Tamil Nadu, India

Department of Chemical Engineering, National Institute of Technology, Tiruchirappalli, Tamil Nadu, India

R. Kabra · S. Singh (✉)
CSB Lab, National Centre for Cell Science, Pune, India
e-mail: singhs@nccs.res.in

S. Singh (ed.), *Machine Learning and Systems Biology in Genomics and Health*,
https://doi.org/10.1007/978-981-16-5993-5_1

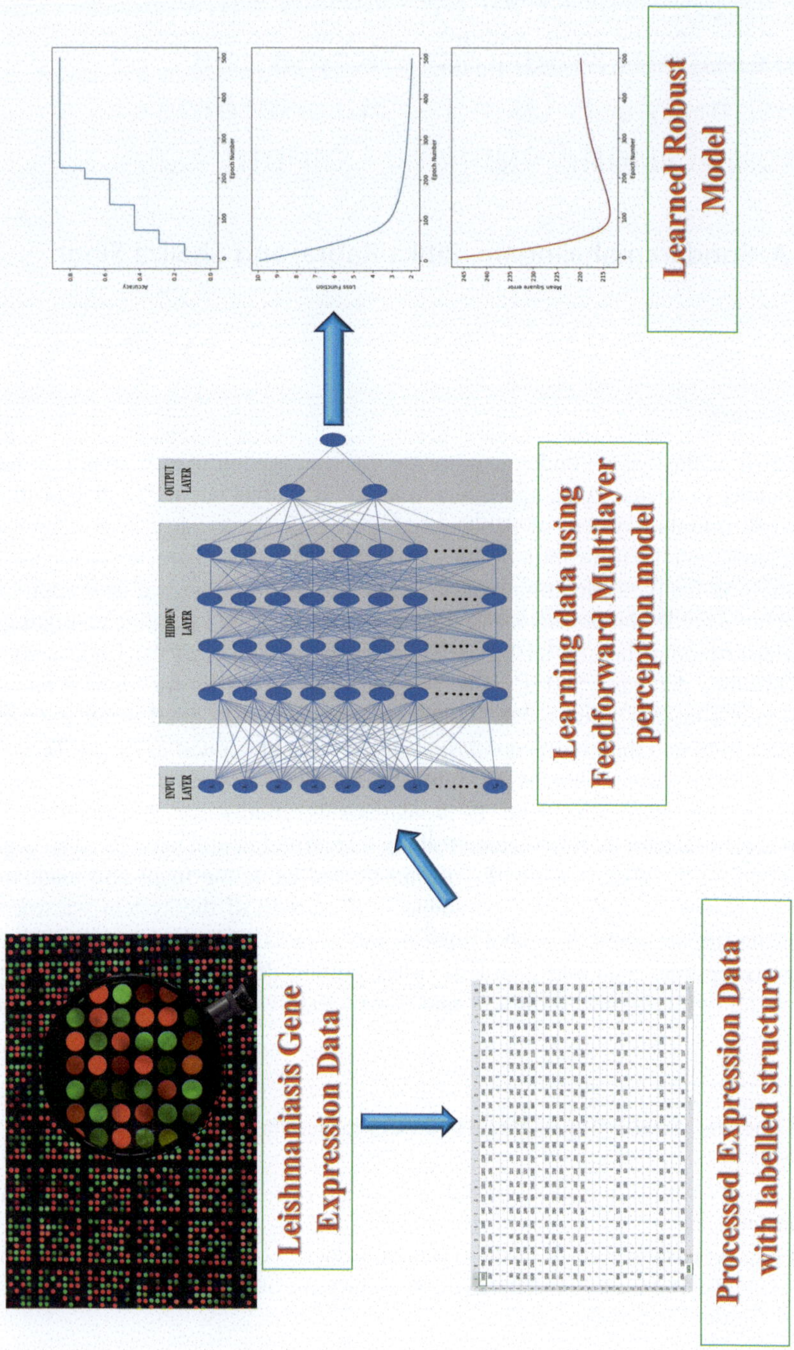
Leishmaniasis Gene Expression Data
Processed Expression Data with labelled structure
INPUT LAYER
HIDDEN LAYER
OUTPUT LAYER
Learning data using Feedforward Multilayer perceptron model
Learned Robust Model

feedforward multilayer perceptron model after normalization, mean square error (219.84), loss function (1.94), and accuracy (85.71%) of the model show the excellent fit of the model with the process, and it could be a better solution for diagnosing leishmaniasis in future, using transcriptome data.

1.1 Introduction

Leishmaniasis is a tropical protozoan parasitic skin disease. It is transmitted through the bite of sand flies, *Phlebotomine* sp. (Mandlik et al. 2012). It is found widely in the tropical and subtropical regions of the world (Mandlik and Shinde 2015). About two million new cases were emerging every year, and they also lead to fatal in some cases. It mainly affects the poor locality of Asia, Africa, and America. This is due to population displacement, insufficient nutrients and immunity in human beings, and local pollutions. Out of 200 countries, 97 are endemic to leishmaniasis. Leishmania is a protozoan parasite responsible for leishmaniasis. There are about 20 *Leishmania* sp. causing primary leishmanial infection in human beings (Orgnization 2017). There are three common types of leishmaniasis: cutaneous leishmaniasis, mucocutaneous leishmaniasis, and visceral leishmaniasis (Kala-azar (or) Black fever) (Vineetha Mandlik et al. 2016). Leishmania parasites are capable of modulating the human immune system for their survival. They also modify the self-healing cutaneous lesions to fatal conditions (Mol et al. 2013, 2015). And some species of leishmaniasis are becoming virulent and drug-resistant. These leishmaniasis features made the researchers focus on the comprehensive identification and analysis of its metabolic pathway and genome (Mandlik et al. 2012).

Leishmaniasis is diagnosed by four different methods approved by US Centers for Disease Control and Prevention. They are slide specimen, culture medium, polymerase chain reaction (PCR), and serological testing. The biopsy specimens, dermal pus (or) scraps, and impression smears are tested as slide specimens. The specialized culture medium is also used for the diagnosis of leishmaniasis. The species identification of the leishmaniasis is carried out using PCR with complementary probes. The serum samples were also tested using the rK39 rapid test kit for diagnosing leishmaniasis (Prevention 2016).

The medical data were significantly managed and secured using computers and networking in the past decades. The increase in the number of patient's data leads to the new technology and technique called data mining, helpful in using processed information. Its evolution leads it to the new health informatics field, used in the health sector to diagnose the disease states and maintain its data. These extended applications of data mining and cognitive computing have paved the way for the physician to diagnose the disease quickly and fast, which has reduced the acute side effects on major diseases. They are also cost-effective and more accurate (Alaa et al. 2018). The diagnosis of disease using artificial neural network (ANN) of feedforward backpropagation networks has yielded a positive prediction model. The working of a biological neuron is mimicked using ANN, which processes the distributed data parallelly. It stores and passes the acquired knowledge of data from one neuron to another for responding to the given situation (or) conditions. ANN is used in normalizing the variance of data, which forms a robust linear model of data. ANN is better than the conventional multiple regression model, which cannot associate the nonlinearity of the biological data. ANN can associate the multidimensional nonlinearized data into a robust model using normalization factors (Torgyn Shaikhina 2017).

A recurrent neural network (RNN) is a fully connected network, similar to a feedforward neural network. But the activation of each layer for RNN is dependent on the output of the previous layer. The temporary output memory of the RNN in passing the activation function leads it to be a good fit model for the prediction of robust data. It can also be activated by the sigmoidal activation function, used in normalizing the deviations in data. The input feed data of the RNN is three-dimensional, stating the sample features, sample size, and time series of length (Ekaansh Khosla et al. 2018).

The multidimensional array of the data is said to be tensor. The graphical and mathematical operations over the collected tensors were carried out through TensorFlow. It is an open-source data frame library. It comprises the libraries for various mathematical computations and graphical interpretations. The flow of numerical data over the mathematical operations is studied using pictorial representations. The large-scale heterogeneous distributed array data are quickly processed by the mathematical operations coded as nodes in a processing algorithm. The primary use of TensorFlow is to handle the big data of the dataset into a machine learning algorithm using suitable coding language (Quan Doa and Chaudri 2017).

Early diagnosis of leishmaniasis would reduce the mortality rate and control the infectious stages of the disease. Hence, it needs a fast-diagnosing model for the control of the disease. This work emphasizes the importance of data mining and cognitive computing in the diagnosis of leishmaniasis. We have proposed a medical diagnostic scheme for diagnosing leishmaniasis using transcriptome data of leishmaniasis. The transcriptome data was processed using RNN. RNN recognizes the variance in leishmaniasis' transcriptome data and forms a robust dataset defining model using the normalization process. This trained model will be applied for the diagnosis of leishmaniasis using transcriptome data.

1.2 Methodology

The varying factors (features) are essential to study the pattern of the given dataset accurately. The transcriptome data of leishmaniasis were collected from the public domains. All the records of the domains were experimentally derived data. The variabilities of the data were used as features and attributes of the algorithm, used in constructing a model for leishmanial diagnosis. The workflow with the leishmaniasis transcriptome data to produce the robust model, helpful in diagnosing leishmaniasis, was shown in Fig. 1.1.

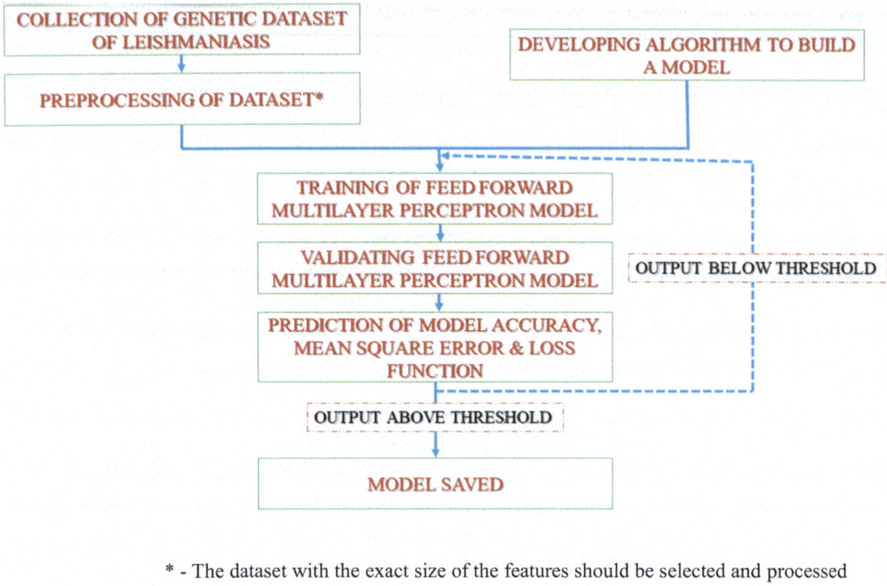

Fig. 1.1 Schematic representation of the progress of workflow. *: The dataset with the exact size of the features should be selected and processed

1.2.1 Collection and Processing of Transcriptome Dataset

The raw transcriptome data were collected from the Gene Expression Omnibus (GEO) database. It is one of the public data repositories of functional genomic data, mainly gene expression data. It comprises the array and sequence-based experimental gene expression data of all diseases. The 33 transcriptome data for total RNA and double-stranded cDNA concentration of the leishmaniasis in human beings with the same number of variables (45,033 rows) were done at the same gene expression array analyzer collected. It comprises five healthy person datasets and 28 infected human datasets. We labelled each dataset with the specific label code and compiled it into a single large dataset in comma-separated values file (.CSV) format. We used 80% of datasets as training data and 20% as the testing data for creating a model. It was further taken for a learning process through the feedforward multilayer perceptron model, using TensorFlow data frame with the developed python code.

1.2.2 Developing a Recurrent Neural Network

A feedforward ANN with multilayer perceptron is used as the biomimic neural network in this study to develop a model for diagnosing leishmaniasis using transcriptome data. We have used TensorFlow 1.8.0 along with the python 3.6 coding platform in this work.

Every node of the ANN was used as the neurons of the human brain. The weights and biases of every node in the first layer were given with the input data. Then the output of every layer was passed as input for the successive layers. We have used four hidden layers in this work. The multilayer perceptron involved in this process was used to study the nonlinear functions of the given genetic data. Backpropagation of data was also done using the code to increase prediction accuracy using the model; it is also named supervised learning of data. It compares the labelled data and the output data for its accurate prediction. The rise or fall in the value was adjusted with the backpropagation process of ANN until it reaches the threshold value of the process. This gradient descent optimization with the backpropagation process minimizes the error. It is also efficient in finding the highly reproducible pattern of the given datasets (Quan Doa and Chaudri 2017).

1.2.2.1 Feedforward Multilayer Perceptron Model

The total transcriptome dataset was split into training (80%) and testing (20%) data. The model was trained at the learning rate of 0.001. The training was repeated for 500 epochs. We processed the data through multilayer perceptron having four hidden layers with 45 neurons in each layer. The accuracy, cost (or) loss function, and mean square error of the model were studied for each iteration and plotted as a graph. The variations in the data were normalized using log2 normalizations. Artificial neural networks could not capture highly dependent variables of the transcriptome data during backpropagation. The gradient of the error function is decreased exponentially with the number of epochs. Using gradient descent optimization algorithm, the highly reproducible data were captured by long short term memory of the model (Kang 2018). The method was followed with the proposed method of Alaa et al. (2018), with slight modification to the operating parameters and hidden layer (Alaa et al. 2018). The model was developed with 45 neurons. It has four hidden layers fully interconnected to each other, classifying two different classes of leishmaniasis. The schematic representation of the working model was given in Fig. 1.2.

1.2.2.2 Calculation of Fitness of the Model

The model was tested for its fitness by calculating the accuracy, loss function, and mean square error of the model at every iterative. The pre-mentioned parameters are calculated using the following equations.

Weight The output of each neuron depends on the weight matrix. The output of each neuron can be given as Eq. (1.1):

$$[Y] = \sum_{i=0}^{n} [x] * [w] \tag{1.1}$$

where $[Y]$ is the output matrix of the corresponding layer, $[x]$ is the input matrix, and $[w]$ is the weight matrix. Neural networks optimize the model learning by changing

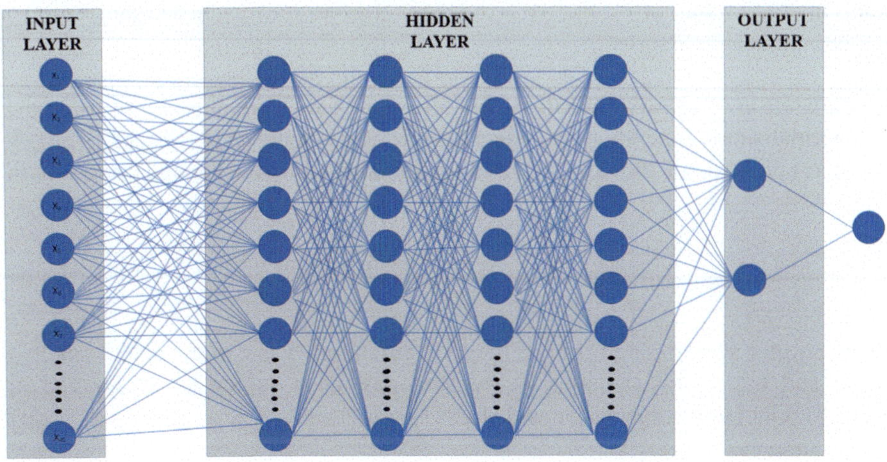

Fig. 1.2 Schematic representation of the feedforward multilayer perceptron model

the weight matrix values between the layer to attain the predefined threshold value. The iterative change in the weight value was carried out through the backpropagation concerning the defined dataset and threshold value (Pizzi et al. 1995).

Layer Output The output of each layer is the sum of the multiplied matrix of input and weight matrix with the bias matrix, given in Eq. (1.2):

$$[Z] = [Y] + [B] \tag{1.2}$$

where $[Z]$ is the output of the layer and $[B]$ is the bias matrix. The extra matrix of the neural network lied before all input layers as a pre-output layer, which is added to the matrix to get the final output of the layer which is termed as bias matrix (Alaa et al. 2018; Pizzi et al. 1995; Payal Dande 2018).

Accuracy The model's accuracy is the ratio of the total true predictions to the overall predictions (Alaa et al. 2018). The accuracy of the model can be calculated by using Eq. (1.3):

$$\text{Accuracy} = \frac{\text{True Positive} + \text{True Negative}}{\text{True Positive} + \text{True Negative} + \text{False Positive} + \text{False Negative}} \tag{1.3}$$

True positive is the number of leishmaniasis datasets correctly classified as leishmaniasis; true negative is the number of datasets of average healthy persons identified as normal. False positive is the number of average person dataset identified as leishmaniasis, and false negative is the number of leishmaniasis dataset identified as a normal healthy dataset.

Normalization The variations in the transcriptome data were normalized using the log2 normalization function by leaving the least reproducing data (Alaa et al. 2018). The normalization of the data process was mathematically denoted as given in Eq. (1.4):

$$\text{Normalized value} = \frac{x\ \text{value} - \text{minimum}\ x\ \text{value}}{\text{maximum}\ x\ \text{value} - \text{minimum}\ x\ \text{value}} \tag{1.4}$$

Mean Square Error The mean of the total variations between the actual labelled output and the predicted output is measured as mean square error, denoted in Eq. (1.5) (Gholamhossein Eslamizadeh 2017):

$$\text{Mean Square Error} = \frac{1}{n} * \sum_{\text{i}=0}^{n} (y_{\text{out}} - y_{\text{actual}})^2 \tag{1.5}$$

Cost (or) Loss Function The model's loss function (or) cost is calculated by the prediction of deviations in the mean squared error between the actual and predicted output value (Torgyn Shaikhina 2017).

1.3 Results and Discussion

1.3.1 Feedforward Multilayer Perceptron Model

The backpropagation-induced feedforward network with multilayer perceptron is activated using the sigmoid function. The usage of backpropagation in the network has decreased the loss function to the prescribed threshold value by adjusting the weights of the neuron connections. The backpropagation incorporated model with multilayer perceptron has added a good solution for the specific problem, diagnosis of leishmaniasis using its transcriptome data. The lower learning rate of the process with the relatively high accuracy of the solution makes it a good model for diagnosing leishmaniasis in human beings using the transcriptome data. The robust and stable model of the process has been developed using backpropagation, which trains nonlinear data into a trained model with mathematical notations. The accuracy, loss function, and the mean square error of the corresponding epoch are tabulated in Table 1.1.

These characters of optimization were explained by Jin et al. (2018). The gradient decrease in the loss function and mean square error concerning their corresponding epoch number is due to the gradient descent optimizer (Kim et al. 2018). This has also increased the accuracy of the model in the prediction of leishmaniasis. The 45,033 features of the individual sample were loaded into the 45 neurons of the input

Table 1.1 Accuracy, loss function, and mean square error of training epochs

Epoch no.	Accuracy (%)	Loss function	Mean square error
0	0	9.91	245.96
25	14.29	7.46	229.82
50	28.57	5.37	219.81
75	42.86	3.86	214.99
100	42.86	3.17	213.62
125	42.86	2.86	213.57
150	57.14	2.67	213.98
175	57.14	2.53	214.58
200	71.43	2.43	215.25
225	71.43	2.36	215.94
250	85.71	2.29	216.60
275	85.71	2.24	217.20
300	85.71	2.19	217.74
325	85.71	2.15	218.21
350	85.71	2.12	218.60
375	85.71	2.08	218.94
400	85.71	2.05	219.21
425	85.71	2.02	219.41
450	85.71	1.99	219.60
475	85.71	1.96	219.74
500	85.71	1.94	219.84

layer. The datasets of 33 different human samples were processed into the input layer. The processed input is passed into the series of hidden layers for its data optimization to select the maximum reproducing data of the given transcriptome data. It was then characterized into the two classes in the output layer. The value of the output class crossing above the threshold value is presented as an output of the given test data. The random structure of the hidden layer used in constructing the model increases the model's efficacy, proving the model's increased accuracy. The randomization of weight input with the hidden layers without affecting convergence would improve the model's fit with the process (Lv et al. 2018). The number of hidden layers required for the model plays a vital role in optimizing the model to act as an efficient model. The usage of the hidden matrix, bias, made the perfect stable model without affecting its convergence (Isabekov and Erzin 2018). The increased accuracy of the model is due to the use of increased neurons in the hidden layer. The increased number of the hidden layer decreases the model's error to the process (Guliyev and Ismailov 2018). The input of each hidden layer is dependent on the output of the preceding neural layer. This dependency structure of the matrix made the model learn the highest data volume of the transcriptome data and also increases the cognitive effect of the model (De Smedt et al. 2018).

1.3.2 Fitness of the Model

1.3.2.1 Accuracy

The accuracy of the model was initially 0%. Then it started learning slowly with the increase in the number of iterations with epochs. They increase as a step-by-step impulse by learning and normalization of data occurring at each stage of the process. Then the absolute accuracy of the data was attained to be 85.71%; they showed it to be a good fit of the model with the process. The variations in the accuracy through learning with their corresponding epoch (or) cycle are denoted in Fig. 1.3. The accuracy of 85.71% shows it to be a good model for predicting the more reliable experimental result for the given transcriptome data of leishmaniasis in human beings. The increase in the accuracy states the increase in controlling the parameters and features with the prediction (Wen et al. 2018). The increasing value of the accuracy denotes the decreasing state of the error in the model, with the increase in the regression coefficient (Huang and Wang 2018). The usage of backpropagation has increased the learning of the model, which in turn increased the accuracy of the model. The usage of backpropagation along the feedforward neural network has also proved to be the better option than using the spike-timing-dependent plasticity model for biological datasets. The deep networks' lower classification accuracy and training difficulty are overcome with this feedforward multilayer perceptron model with backpropagation (Kim et al. 2018). An accuracy value above 80% is accepted since it could classify and identify the class with its unique features with 70% of total accuracy (Kusy and Kowalski 2018).

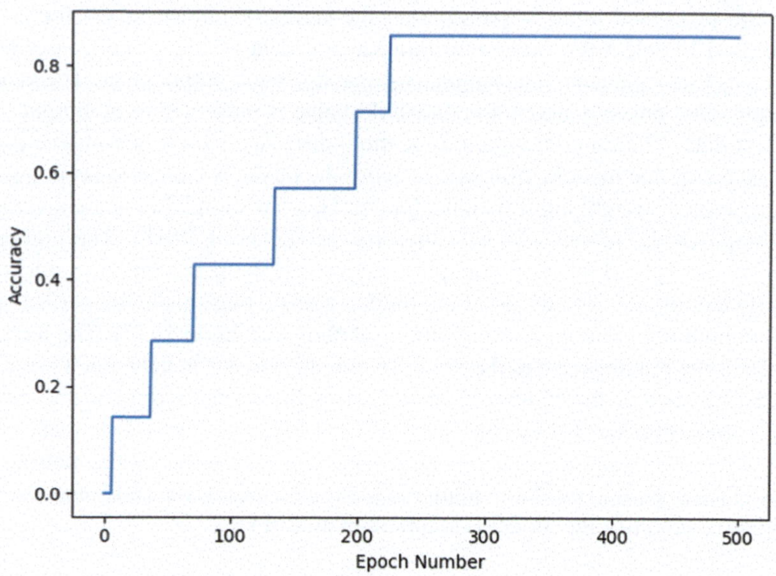

Fig. 1.3 Effect of accuracy with their corresponding epoch number

1.3.2.2 Loss Function

The loss function (or) cost of the model shows a significant decrease in the value with the continuous increase in epochs. The loss function was found to be 9.91 at the start of the training process. Then the rise in the number of cycles (or) epochs decreases the loss function (or) cost, with continuous learning through the datasets repeatedly. The further increase in the number of cycles used for training the model using the training dataset has decreased the model's loss function (or) cost. The variations of the loss function with their corresponding epoch are given in Fig. 1.4.

The intrusion of loss function into the model depicts the minor change of the dataset regarding its variations. It is also used in studying the impact of improvement. The improvement in the loss function shows the method's efficiency in identifying the relatively indistinguishable data (Qian et al. 2018a). The matching process of pairs is higher with the decrease in the loss function. The decrease in loss function states the excellent fit of parameters with the developed model (Gu et al. 2018). The stability of the loss function shows the strength of the model with the given data. It also offers the improved functionality of parameters with reduced error (Takase et al. 2018). The decrease in the loss function shows the trained model with higher accuracy and proves the backpropagation's ideological property in learning. It is also used in establishing the combination of parameter coefficients to be less with its improved values (Qian et al. 2018b).

1.3.2.3 Mean Square Error

The mean square error of the process was found to be 245.96 initially. There is a gradual decrease in the mean square error value to 100 epochs. Then there is a slight

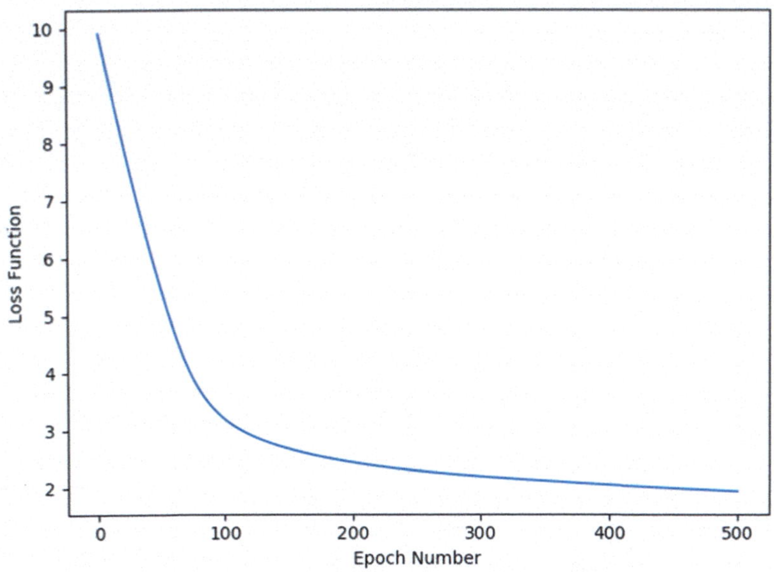

Fig. 1.4 Effect of loss function with their corresponding epoch number

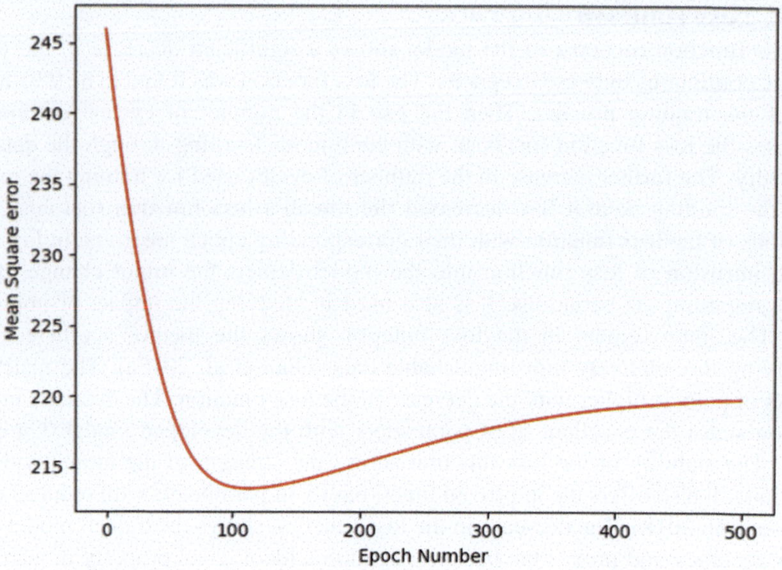

Fig. 1.5 Effect of mean square error with their corresponding epoch number

increase in the value from 100 to 300 generations. Then it was standardized to a constant value of 219 after 300 generations, which shows the normalized state of all data. The learning process with the leishmaniasis transcriptome data was normalized to the yield of the robust model, leaving out the least reproducing features. The process of normalization was carried out in learning the dataset using the log2 normalization method. The variations in the mean square error with their corresponding epoch number were plotted in Fig. 1.5. The smaller decrease of the mean square error is because of the high nonuniformity of the given dataset. The maintained level of deviations in the plot proves it to be normalized from higher variations of data (Ryczko et al. 2018). The higher mean square error value is due to the lower samples of data (Durodola et al. 2018). The mean square error value decrease also proves the working of the algorithm's feedback gradient descent optimization mechanism. The parameters are made to fit into the model correctly by improving the weight of the input data (Rady 2011). It also proves the convergence of the data reliable to the robust model for the given dataset for the constant parameters (Mohammadi et al. 2018). The mean square error is also used in studying the performance of the model with the neural network. The decrease in mean square error states the reduction in error between the network output and the target output, showing the excellent fit of the model with the process (Katić et al. 2018).

1.3.3 Other Influencing Parameters of the Models' Fitness

Epoch The classification (or) diagnosis is not efficient at the lower epochs, whereas on continuous learning through several iterations, the classification efficiency is increased. These epochs are the responsible parameter of the accuracy and time definer of the developed model (Balaji et al. 2021; Zhang et al. 2021).

Latency and Throughput Latency is the time required by the model to convert the input values into a final output. Throughput is the pack volume processed per unit time. This latency and throughput value affect the efficacy of the model (Mittal and Umesh 2021).

Learning Rate and Optimizer The choice of learning rate and optimizer make the model learn with higher accuracy at a lower time limit. It is generally maintained at 0.001 as a learning rate for the model development to meet the higher efficiency in less time (Nasir et al. 2021).

1.4 Conclusion

This study resulted in a new attempt to process the human sample's transcriptome data into the human disease diagnosis using cognitive computing. The hybridization of the transcriptome data with its prediction model, created using the algorithm, is the best solution for diagnosing leishmaniasis in human beings in the future. The use of the feedforward multilayer perceptron model with backpropagation has produced a good fit for diagnosis, showing 85.71% accuracy in prediction. We have brought a new solution for predicting disease in human beings using the transcriptome data. The classification of leishmaniasis in human beings using their transcriptome data of total RNA and c DNA concentration has resulted in the new prediction model. The accuracy, mean square error, and loss function prove the model to be a good fit for diagnosing leishmaniasis using the transcriptome data. So, we suggest this model can diagnose leishmaniasis in human beings using their transcriptome data. This cognitive computing approach in diagnosing leishmaniasis in human beings will also form a robust dataset emphasizing the maximum reproducing data, a target set for the in vitro research regarding leishmaniasis in human beings. In the future, the model would be trained with a large number of the dataset, thereby increasing the model's cognitive level in the diagnosis and prediction of the given dataset. It would also be made as an application for the leishmanial diagnosis in human beings using transcriptome data. This finding could also be a key factor for other diagnosis using the extensive volume data in the column basis learning using RNN.

References

Alaa S, Al Agha HF, Bassam H, Hammo, Ala M, Al-Zoubi (2018) Identifying β-thalassemia carriers using a data mining approach: the case of the Gaza strip. Artificial Intelligence Medicine, Palestine

Balaji E, Brindha D, Elumalai VK, Vikrama R (2021) Automatic and non-invasive Parkinson's disease diagnosis and severity rating using LSTM network. Appl Soft Comput 108:107463

De Smedt J, De Weerdt J, Serral E, Vanthienen J (2018) Discovering hidden dependencies in constraint-based declarative process models for improving understandability. Inf Syst 74:40–52

Durodola JF, Ramachandra S, Gerguri S, Fellows NA (2018) Artificial neural network for random fatigue loading analysis including the effect of mean stress. Int J Fatigue 111:321–332

Ekaansh Khosla DR, Sharma RP, Nyakotey S (2018) RNNs-RT: flood based prediction of human and animal deaths in Bihar using recurrent neural networks and regression techniques. Proc Comp Sci 132:486–497

Gholamhossein Eslamizadeh RB (2017) Heart murmur detection based onWavelet transformation and a synergy between artificial neural network and modified neighbor annealing methods. Artif Intell Med 78:1–29

Gu J, Wang Z, Kuen J, Ma L, Shahroudy A, Shuai B, Liu T, Wang X, Wang G, Cai J, Chen T (2018) Recent advances in convolutional neural networks. Pattern Recogn 77:354–377

Guliyev NJ, Ismailov VE (2018) On the approximation by single hidden layer feedforward neural networks with fixed weights. Neural Netw 98:296–304

Huang L, Wang J (2018) Global crude oil price prediction and synchronization based accuracy evaluation using random wavelet neural network. Energy 151:875–888

Isabekov A, Erzin E (2018) On the importance of hidden bias and hidden entropy in representational efficiency of the Gaussian-bipolar restricted Boltzmann machines. Neural Netw 105:405–418

Kang S (2018) Personalized prediction of drug efficacy for diabetes treatment via patient-level sequential modeling with neural networks. Artif Intell Med 85:1–6

Katić K, Li R, Verhaart J, Zeiler W (2018) Neural network based predictive control of personalized heating systems. Energ Buildings 174:199–213

Kim J, Kim H, Huh S, Lee J, Choi K (2018) Deep neural networks with weighted spikes. Neurocomputing 311:373–386

Kusy M, Kowalski PA (2018) Weighted probabilistic neural network. Inform Sci 430-431:65–76

Lv Z, Zhao J, Zhai Y, Wang W (2018) Non-iterative T–S fuzzy modeling with random hidden-layer structure for BFG pipeline pressure prediction. Control Eng Pract 76:96–103

Mandlik V, Shinde S (2015) Structure based investigation on the binding interaction of transport proteins in leishmaniasis: insights from molecular simulation. Mol Biosyst 11:1251

Mandlik V, Shinde S, Chaudhary A, Singh S (2012) Biological network modeling identifies IPCS in Leishmania as a therapeutic target. Integr Biol 4(9):1130–1142

Mittal S, Umesh S (2021) A survey on hardware accelerators and optimization techniques for RNNs. J Syst Archit 112:101839

Mohammadi M, Tan Y-H, Hofman W, Mousavi SH (2018) A novel one-layer recurrent neural network for the l 1 -regularized least square problem. Neurocomputing

Mol M, Kosey D, Singh S (2015) Nano-synthetic devices in leishmaniasis: a bioinformatics approach. Front Immunol 6(323):1–6

Mol M, Patole MS, Singh S (2013) Immune signal transduction in leishmaniasis from natural to artificial system: role of feedback loop insertion. Biochim Biophys Acta 1840:71–79

Nasir JA, Khan OS, Varlamis I (2021) Fake news detection: a hybrid CNN-RNN based deep learning approach. Int J Inform Manage Data Insights 1(1):100007

Payal Dande PS (2018) Acquaintance to artificial neural networks and use of artificial intelligence as a diagnostic tool for tuberculosis: a review. Tuberculosis 108:1–9

Pizzi N, Choo LP, Mansfield J, Jackson M, Halliday WC, Mantsch HH, Somorjai RL (1995) Neural network classification of infrared spectra of control and Alzheimer's diseased tissue. Artif Intell Med 7:67–79

Prevention, U.C.f.D.C.a (2016) Diagnosis of Leishmaniasis. CDC's Division of Parasitic Diseases and Malaria

Qian S, Liu H, Liu C, Wu S, San Wong H (2018a) Adaptive activation functions in convolutional neural networks. Neurocomputing 272:204–212

Qian S, Liu H, Liu C, Wu S, Wong HS (2018b) Adaptive activation functions in convolutional neural networks. Neurocomputing 272:204–212

Quan Doa TCS, Chaudri J (2017) Classification of asthma severity and medication using tensor flow and multilevel databases. Proc Comput Sci 113:344–351

Rady HAK (2011) Shannon entropy and mean square errors for speeding the convergence of multilayer neural networks: a comparative approach. Egyptian Inform J 12(3):197–209

Ryczko K, Mills K, Luchak I, Homenick C, Tamblyn I (2018) Convolutional neural networks for {Nasir, 2021 #35} atomistic systems. Comput Mater Sci 149:134–142

Takase T, Oyama S, Kurihara M (2018) Effective neural network training with adaptive learning rate based on training loss. Neural Netw 101:68–78

Torgyn Shaikhina NAK (2017) Handling limited datasets with neural networks in medical applications: a small-data approach. Artif Intell Med 75:51–63

Vineetha Mandlik SP, Bopanna R, Basu S, Singh S (2016) Biological activity of Coumarin derivatives as anti-leishmanial agents. PLoS One 11(10):1–15

Wen S, Xie X, Yan Z, Huang T, Zeng Z (2018) General memristor with applications in multilayer neural networks. Neural Netw 103:142–149

World Health Organization (2017) Global Health Observatory (GHO) data, 2016. In: Leishmaniasis: situation and trends. WHO, Geneva

Zhang L, Chen D, Chen P, Li W, Li X (2021) Dual-CNN based multi-modal sleep scoring with temporal correlation driven fine-tuning. Neurocomputing 420:317–328

Big Data in Drug Discovery

2

Sanjay Bhattarai, Rajnish Kumar, Sagorika Nag,
and Vigneshwaran Namasivayam

Abstract

Drug discovery is a challenging and complicated process that requires decades to discover and develop a drug. This process can be streamlined and simplified by using big data. Big data is applied in both chemical and biological aspects of drug discovery starting from target validation to the final stage of clinical trials and explained with an example focusing on medicinal chemistry. This chapter lists the public, commercial, and proprietary chemical space available for drug discovery. The application of AI in drug discovery provides greater support and improves its efficiency. The important softwares, scripts, data analysis, and mining tools available which aid in the identification of a novel drug and development are provided. Among the different stages in drug discovery, hit identification for a therapeutic target is an important and time-consuming phase. This can be handled by using proprietary and commercial screening platforms with the help of big data analysis and management. Thus, the importance of screening platforms and their success stories are highlighted with case studies. Overall, the integration of big

Sanjay Bhattarai, Rajnish Kumar contributed equally to this work

S. Bhattarai
Department of Medicinal Chemistry, School of Pharmacy, University of Kansas, Lawrence, KS, USA

R. Kumar · S. Nag
Department of Pharmaceutical Engineering & Technology, Indian Institute of Technology (B.H. U.), Varanasi, Uttar Pradesh, India

V. Namasivayam (✉)
PharmaCenter Bonn, Pharmaceutical Chemistry, Pharmaceutical Institute, University of Bonn, Bonn, Germany
e-mail: vnamasiv@uni-bonn.de

S. Singh (ed.), *Machine Learning and Systems Biology in Genomics and Health*,
https://doi.org/10.1007/978-981-16-5993-5_2

data with AI-based tools greatly improves the efficiency of the drug discovery process by making it simple and less time-consuming.

2.1 Introduction

2.1.1 What Is Big Data?

Information has become the most important commodity of today's world because of its widespread use in several sectors including the pharmaceutical and biotech industry. The day-to-day activities have resulted in data that is too big, complex, and difficult or impossible to process using classical approaches. The collected information has the capability to be analyzed computationally to reveal patterns, trends, and associations known as "big data." In the last decade, there have been enormous challenges in obtaining data, storage of data, scientific and systematic analysis of data, retrieval of stored data, secured sharing and transfer of data, visualization and proper analysis of data, periodic update of data, as well as information privacy and protection of raw data sources. The development and advancement in information technology allow us to analyze and interpret the collected information and make the best use of it. Big data are highly important as it provides new ways in which data is being stored and is responsible for the huge growth of the pharmaceutical and biotech industry (Zhu 2020; Moingeon 2021).

Big data analytics cover the integration of diverse information and do quality verification, analysis, modeling, elucidation, and validation (Zang et al. 2017). The introduction of big data analytics provides an extensive understanding of the enormous amount of data available. The major characteristics of big data that have been studied vigorously when it comes to the process of drug discovery and development are volume, velocity, and variety. The other two characteristics that are highly relevant in the context of pharmaceutical research and development are veracity and value. All these five characteristics make up the 5 V's of big data (Reimer and Madigan 2018; Kitchin and McArdle 2016).

Volume: The volume of big data is one of its most important attributes. It is usually considered by taking into account its expansion or compression after the processing stage in comparison to the original raw data. Choosing the potential molecule as a drug for a specific target from a large set of molecules is an example of how the volume of big data plays a vital role in the pharmaceutical and biotech industries (Archenaa and Anita 2015).

Variety: Variety of big data refers to the format of raw data available. The process is considered easier if the data is available in a structured format. However, most of the data available now are present in a semi-structured or unstructured format which requires an additional preprocessing step in the data collection.

Velocity: The speed at which new information is made available is called velocity. The approaches used to process the data significantly depend on velocity. An example of static data is whole-genome sequence information of a patient at a specified period of time and screening thousands of compounds in the high-throughput screening (HTS) platform in a couple of hours, or computational screening of a million subsets of compounds against a set of target within an hour. The latter example would require real-time analysis of the data and complex methodologies in comparison to the prior one.

Veracity: It refers to the accuracy and truthfulness of the data that is to be processed (Reimer and Madigan 2018). As said by George Fuechsel, an early IBM programmer, "garbage in, garbage out" means that, if the quality of input data is not good, then the resulted data will not achieve good results (Kilkenny and Robinson 2018). Especially in life sciences, where data contributes to the wellness and health of people, the veracity aspect of data is extremely important.

Value: The value of the collected information as a dataset should be measured from the beginning and it has to be further justified for the effort involved in the processing and analyzing of the data.

Due to the huge resources of data available carrying the information about pharmacokinetic (PK) and pharmacodynamics (PD) properties of chemical compounds, the pathophysiology of diseases, design of investigation trials, patients selected for the trials and their disease conditions, the process of drug discovery and development has become more efficient and safer. Computational methods now are being used not only to discover lead compounds but also to predict the required pharmacological attributes before they reach clinical trials (Ristevski and Chen 2018).

Besides this, big data allows us to optimize and increase the efficacy of clinical trials, identification of target-specific patient populations with desired effectiveness to get a better insight into patient behavior for enhancement in drug delivery, advancement in safety with better risk management, and increased marketing and sales performance of drug molecules. Based on the application of big data in the pharmaceutical and biotech industry (Luo et al. 2016; Agrawal and Prabakaran 2020), it can be stated that big data is about better understanding of the available sources, but the challenge appears in the way it needs to be processed and to achieve maximum benefit from it.

2.1.2 Drug Discovery

Drug discovery is generally explained as the process of finding bioactive compounds for the development of novel drugs and is one of the first stages in a multistage drug discovery and development pipeline (Chen et al. 2018a). The drug discovery process is very time-consuming, expensive, and carries a high risk of failure (Mohs and Greig 2017). A drug discovery program initiates due to a specific sickness, disease, or chronic problem without adequate drugs available for treatment, and this unmet clinical requirement is the driving force for the drug discovery projects. The basic

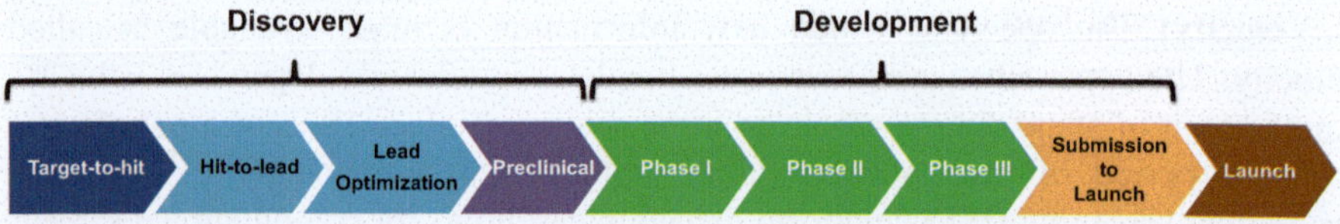

Fig. 2.1 Drug discovery and development process. The first step in drug discovery is the identification and validation of a target for a particular disease. High-throughput assays are then designed on the base of the target. The screening of compound libraries by HTS results in HIT molecules. Further, the molecules are optimized with respect to potency, physiochemical properties: drug-likeness, SAR studies, detail ADMET, PK/PD, resulting in preclinical candidates. Preclinical candidates are then investigated in the in vivo animal disease model to observe the desired pharmacological effect resulting in clinical candidates. After getting the regulatory approval, the clinical trials (Phase I-III) are conducted, and if successful, Food and Drug Administration (FDA) can approve for the commercial use of the clinical candidates as drugs

research of the drug discovery process is often initiated by academia, and they generate data for the development of a hypothesis that the inhibition or activation of a protein/pathway will result in a therapeutic effect of a disease condition. The outcome of this activity results in the selection of a pharmacological target and its validation. The industries take a principal role in lead discovery effort, which is an intensive, vigorous, and robust search of drug-like small molecules or biological therapeutics, typically termed a development candidate or new chemical entities (NCEs). In the later stage of drug discovery, NCEs progresses into preclinical and, if successful, finally into the clinical trials and ultimately become a marketed drug.

An overview of the drug discovery and development process is shown in Fig. 2.1. Drug discovery mainly includes four important stages: (a) target identification and validation, (b) model establishment for the high-throughput screening (HTS) of chemical compounds, (c) identification of hits and optimization of hit molecules resulting in the discovery of lead compounds, and (d) optimization of the lead compounds with the preclinical studies resulting in NCEs or clinical candidates. Once these processes are completed, the clinical trials (Phase I–III) are initiated (Dickson and Gagnon 2004; Paul et al. 2010).

1. **Target identification and validation**: It is the initial step in the drug discovery process for a particular disease. It consists of identification and validation of the therapeutic target which is usually a gene or a protein and to understand its role in the disease (Kopec et al. 2005). This is followed by the characterization of the molecular mechanism of the validated target.
2. **Model establishment**: After the target selection, the experimental studies including in vitro assays, in vivo assays, binding affinity studies, kinetic studies, gene knockout studies, and studies on transgenic animal models are required to screen and estimate the biological and pharmacological activity of the potential drug candidates. The experimental models are developed or adapted according to the screening criteria and test molecules or other drug candidates. This helps to determine whether these molecules or drug candidates satisfy the requirements of the action and to ensure the specificity of the research process.

3. **Discovery of lead compounds**: To identify a lead compound that could target a unique gene, a specified protein, or a signaling pathway, thousands of small molecules are evaluated in laboratories with a variety of strategies and by applying different experimental methods and techniques. There are many ways to obtain new lead molecules, which are identified from extensive experimental screening such as characterizing natural products, combinatorial library screening, and computational-based approaches such as structure- or ligand-based virtual screening through a rational drug design approach (Chen et al. 2018a).
4. **Optimization of the lead compounds:** The optimization of a lead compound is one of the most important steps in the drug development process. Once an initial lead compound is identified, optimization will be applied to further improve potency, selectivity, toxicity, safety, molecular mechanism, and distribution. The lead compound may have issues such as low action intensity or specificity, inadequate pharmacokinetic properties, toxicity or strong side effects, and chemical or metabolic instability. This compound has to be optimized before it has to be considered as a candidate molecule for preclinical, clinical studies, and later as a drug.

2.2 Big Data in Drug Discovery

2.2.1 Big Data in Chemistry

In medicinal chemistry, the data-driven approaches have the ability to improve the decision-making steps and enhance the quality of the drug discovery projects. The availability of big data handling tools for researchers helps to reveal the interesting relationships and patterns in available data. With the developments in science and technology, there has been a substantial improvement in the quality of research. In the current multidisciplinary research, the increasing amount of data generation offers incredible possibilities for data-driven science, with the potential to transform the existing activities. The application of so-called "big data" offers to undertake research projects which were never possible before and encourages to re-evaluate the scientific practices using the available information. The recently developed artificial intelligence (AI) approaches using machine learning and deep learning algorithms and relevant molecular modeling studies based on big data offer new solutions to efficacy and safety evaluations of drug candidates. The generated models provide deep insights from the chemical structure of the molecules to in vitro, in vivo, and clinical studies. The appropriate data mining, curation, and management approach provide great support to recent studies (Lusher et al. 2014). High-throughput screening resulting in hit generation and chemical synthesis for optimization of leads are the two highly time-consuming and costly processes in drug discovery. These challenges can be addressed by turning to alternate strategies where computer-based screening, for example, virtual screening, often substitutes for the expensive and time-consuming high-throughput screening and the combinatorial chemistry which ensures the synthesis of a large number of chemical compounds in a single process which is fast, cheap, and comparatively easy. Though both in silico screening and combinatorial chemistry seek to make drug discovery more feasible, it is not always practical to couple these two methodologies very easily (Zhu 2020).

In general, big data in chemistry has accelerated the drug discovery process. The big data related to available chemical spaces, libraries, databases, and a variety of huge resources has allowed researchers to access billions of compounds from all parts of the world, reducing the overall cost risk as well as increasing the effectiveness of the drug.

2.2.1.1 Chemical Data and Physicochemical Properties

In computer-aided drug discovery, the use of large chemical libraries is very common, and assessing their overlap/complementarity is a routine task. For this purpose, a variety of methods such as similarity searching based on fingerprints or molecular descriptors and a comparison of physicochemical properties are available. However, these methods are applicable only if the compound datasets are small, and for chemical spaces containing millions or billions of compounds (big data), it requires huge computational power. Therefore, it is required to apply other faster methods of assessment while working with chemical spaces containing billions of compounds.

The advancement in technology along with the huge volume and variety of data available has allowed researchers to come up with platforms that assist to visualize the process of drug discovery and innovative research. The platforms from proprietary software such as Schrödinger (https://www.schrodinger.com/), BioSolveIT (https://www.biosolveit.de/), and Molecular Operating Environment (MOE) (https://www.chemcomp.com/) aid researchers in structure- and ligand-based drug discovery. The visualizations provided by such software make the computation work behind the drug discovery, very easy. The modeling platforms help in interactive, visual, compound prioritization along with compound evolution. The virtual screening approach uses the basis of similarity, to find molecules of interest in all the chemical spaces available.

The behavior of drugs in humans generally depends on physicochemical properties, such as octanol-water partition coefficient (log P), water solubility (log S), melting point, boiling point, protein binding, and hydrogen bonding. These properties affect the bioavailability, permeability, absorption, transport, and metabolism of drugs in the body and are used extensively in exposure, toxicological hazard, and risk assessments of drugs (Gombar et al. 2003). With the known information available from different experiments, it provides an opportunity to use and generate models for predicting these physicochemical or pharmacokinetic properties. Different models by applying AI are now available for choosing the best-fitted compound with the best set of these properties (Cáceres et al. 2020; Kar and Leszczynski 2020).

2.2.2 Big Data in Biology

The recent advancement in screening technologies has accelerated the amounts of omics data including epigenome, transcriptome, proteome, and metabolome. Earlier the data from individual sources (e.g., genome) were analyzed using machine learning and statistical tools. However, the current practice of integrative analysis of the multi-omics and clinical data helps in several biomedical discoveries and achieves great advancement in pharmaceutical and biotechnological research (Hulsen et al. 2019). This integration has posed several new challenges in addition to the problems associated with single-omics studies. For dealing with these problems, special computational methods are required in order to increase the effectiveness and efficiency of integrative analysis of data acquired from different disciplines (Mirza et al. 2019).

2.2.2.1 Target Identification and Validation, Genomics, Proteomics, and Drug Repurposing

The target identification and validation is the initial step in the drug discovery process and considers a specific target for the development of a drug (Kopec et al. 2005). It evaluates the selected molecular target that could benefit from the drug to be identified and for therapeutic application (Blake 2007). The application of computer-aided approaches in target identification and validation based on the available data could assist in understanding the complex biological pathways and enhance the discovery and development process. Especially, the data for drug-target interactions is given special importance in computational biology, which could provide a better understanding in the process of identifying new candidate molecules for biological validation. Furthermore, the approaches can speed up the discovery and reduce the cost of the experiments. Target validation ensures that the target has potential therapeutic benefits which are an important prerequisite for drug development projects. In the absence of a valid target, a drug discovery program might result in failure (Blake 2007).

Genomics can be defined as the systematic analysis of genes, their roles, and their relationships, and proteomics as the systematic analysis of proteins, protein complexes, interactions, and post translational modification. Till few years back, genomics and proteomics research studies were focused on a single gene or protein at a time. With the recent advancement in high-throughput technologies, there is a paradigm shift from a traditional hypothesis-driven to a data-driven approach. A large number of genes and proteins can be measured and their activity and interaction can be calculated which contributes to find new insights into mechanisms of living systems (Frantzi et al. 2019; Russell et al. 2013).

Drug repurposing is a concept of finding new indications for already known approved drugs. The approved drugs act as an effective source with the availability of big data from chemistry, genomics, biology, and clinical trials (Talevi and Bellera 2020). The detailed investigation of the information available from the approved drugs helps to design an effective drug for different therapeutical targets. It emerged as an alternative approach for rapid identification of drugs for a rare and complex disorder for which an effective drug is not available (Glicksberg et al. 2019). Recently, for the treatment of COVID-19 different approved drugs are explored as an initial treatment to combat the disease.

2.2.3 Big Data in the Pharmaceutical Research Focusing Medicinal Chemistry

In 2011, the Food and Drug Administration (FDA) approved the drug crizotinib (Xalkori) developed by Pfizer targeting non-small cell lung cancer caused due to the rearrangement of the anaplastic lymphoma kinase (ALK) gene (Fig. 2.2). The lead molecule of the crizotinib, PHA-665752, is a potent mesenchymal-epithelial transition (MET) receptor tyrosine kinase inhibitor; however, it has issues in physicochemical properties including pH-dependent low solubility, high metabolic clearance, and poor membrane permeability (Cui et al. 2011; Poon and Kelly 2017). The molecule was optimized by applying the structure-based drug design approach using the co-crystal structure of PHA-665752 in complex with the MET kinase domain to have less lipophilic and as a small hinge binder for better interaction in the pocket of the kinase. The optimization and evaluation in a panel of 120 kinases found crizotinib as potent and selective for MET and ALK (Shaw et al. 2011). The development of crizotinib by Pfizer is mainly based on the application of three different platforms, the genomics research repository, the life sciences hub, and the repository of electronic medical records. The genomics research repository is an open-source data management system from Johnson & Johnson and Recombinant Data Corp. The life sciences hub was initially built to store data from clinical trials research by Pfizer and its CRO partners. The repository for electronic medical records, originally built on Teradata a storehouse of millions of anonymized records. The support of these three platforms along with their data analysis has made it possible by optimizing the lead molecule, PHA-665752, to develop the precision medicine, crizotinib (Vuong et al. 2020).

2.3 Big Data Resources

2.3.1 Public Resources

In recent years there has been a huge increase in the availability of data containing binding affinity, functional activity, ADMET, structures of the protein and ligand and makes it possible to explore the information and apply it to solve a research problem. In particular, the information available in public resources is tremendously increasing in addition to the chemical structural information of the molecules, and here few of the resources are listed.

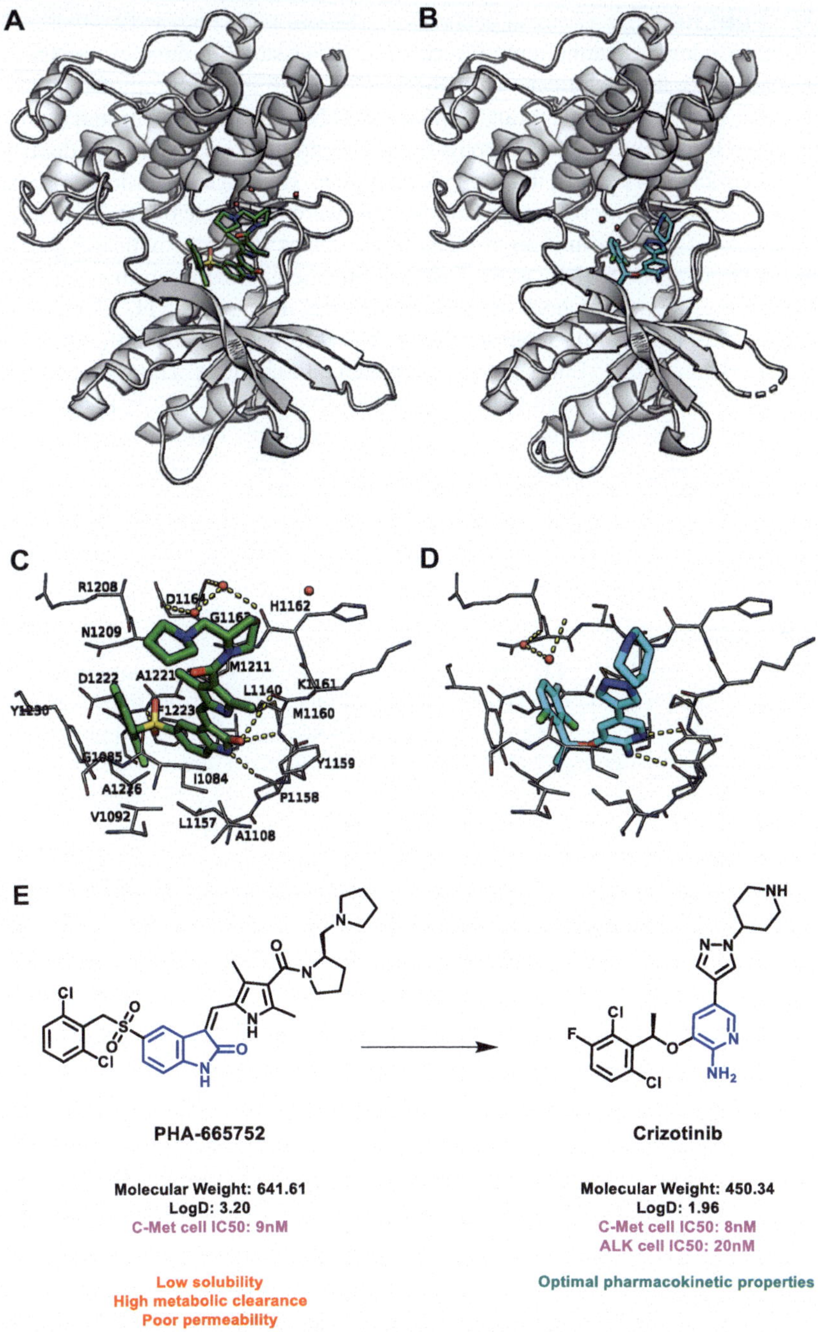

Fig. 2.2 Discovery of crizotinib (Xalkori) from PHA-665752. The crystal structures of c-Met in complex with (**a**) PHA-665752 (green, PDB ID: 2WGJ) and (**b**) crizotinib (cyan, PDB ID: 2WKM). The binding pocket of c-Met and amino acid interaction with the (**c**) PHA-665752 and (**d**) crizotinib are shown. The c-Met kinase is shown in cartoon representation (gray), important amino acids in the binding pocket as line representation (gray), and water molecules in the sphere model. The oxygen

2.3.1.1 PubChem

PubChem is a database containing information about small molecules, nucleotides, lipids, carbohydrates, peptides, and chemically modified macromolecules. It is maintained by the National Library of Medicine (NLM) and the National Institutes of Health (NIH). PubChem curates information on chemical structure, identifiers, physicochemical properties, biological activities, safety, and toxicity. Besides, World Intellectual Property Organization (WIPO) collaborates with PubChem to search a dataset of >16 million chemical substances through its patent database called PATENTSCOPE (Kim et al. 2020). After the discontinuation of TOXNET databases in 2019, PubChem is hosting several data from TOXNET which includes Chemical Carcinogenesis Research Information System, Hazardous Substances Data Bank (Fonger 1995), Genetic Toxicology Data Bank (Gene-Tox) (Auletta et al. 1991), and Liver Tox (Hoofnagle et al. 2013) for providing toxicological information about the chemical substances (Kim et al. 2020). The information available in PubChem serves as a big data resource for creating models using AI and data analysis which includes virtual screening, drug repurposing, prediction of chemical toxicity, drug side effects, and identification of metabolites.

2.3.1.2 ChEMBL

ChEMBL is a bioactivity database containing information about small molecule bioactivity curated from medicinal chemistry journals and data from approved drugs. In order to make it more comprehensive, the bioactivity data is exchanged with PubChem BioAssay (Wang et al. 2017) and BindingDB (Gilson et al. 2016). The analysis of data is an important part of database management; therefore, the information available for compounds, assay, and biological activity is assessed manually with the support of analytical tools to avoid errors and enhance the reliability of the data. The latest release of ChEMBl28 contains approximately 2.6 million compounds, 17.3 million activities, 1.4 million assays, and 14,347 targets. During the recent COVID-19 pandemic, to help the scientific community and accelerate the drug discovery of antiviral drugs, ChEMBL released a database containing curated information about the bioactivity of cell-based assay results of several screened libraries (ChEMBL_27-sars-cov-2-release [Available from: http://chembl.blogspot.com/2020/05/chembl27-sars-cov-2-release.html]).

ChEMBL contains three thematic portals, ChEMBL Neglected Tropical Disease (NTD) as a repository dedicated to neglected tropical diseases, UniChem that produces cross-references linking chemical structure identifiers from different databases, and SureChEMBL which collects information of compounds from patent

Fig. 2.2 (continued) atoms are colored in red, nitrogen in blue, sulfur atoms in yellow, chlorine atoms in light green, and fluorine atoms in light cyan. (**e**) The 2D representation of the structures with the hinge binder indolin-2-one in PHA-665752 is replaced by 2-aminopyridine in crizotinib colored in blue, and the physicochemical properties and biological activities of the molecules are provided

documents (Mendez et al. 2018). Moreover, it provides a malaria inhibitor prediction platform that consists of a consensus model for predicting blood-stage malaria inhibition to predict the antimalarial activity of new molecules (Bosc et al. 2020).

2.3.1.3 BindingDB

BindingDB consists of approximately 1,132,739 experimentally determined protein-ligand binding affinities data along with 489,416 small molecules and 7020 protein targets (Gilson et al. 2016). It curates the data collected from the PubChem (Kim et al. 2020), ChEMBL (Gaulton et al. 2012), Psychoactive Drug Screening Program (PDSP) Ki (Roth et al. 2000), and CSAR databases (Dunbar Jr. et al. 2013), in order to integrate the protein-ligand affinity data as a unified interface in BindingDB. One of the features of BindingDB is that it acts as a warehouse and contains detailed experimental information like pH and temperature of the assay performed to evaluate the compounds. For data curation, it uses a combination of automated and manual procedures to import only those measurements that have a well-defined protein target and a quantitative measure of affinity or relative affinity, usually expressed as IC_{50}, K_i, or K_d value, and to elaborate entries as required to stick to its stringent specifications which make this data highly useful for medicinal chemists and chemical biologists working on the development of structure–activity relationship and QSAR for lead optimization and preclinical drug development projects (Gilson et al. 2016).

2.3.1.4 ZINC Database

The selection of compounds after virtual screening and their availability to procure them for biological testing in different screening assays is an important part of hit identification. For this, ZINC hosts large libraries of small molecules curated from different vendors, pharmaceutical companies, proprietary libraries, and clinical drugs. Overall, the database contains over 120 million purchasable compounds and drug-like compounds with almost 25 percent of that is ready to deliver (Sterling and Irwin 2015). The primary aim of the ZINC database is to provide compounds for virtual screening; however, in the update release ZINC 15, it has become one of the major tools for ligand discovery by connecting molecules with their biological activity and commercial availability (Sterling and Irwin 2015). The compounds available in the ZINC database are classified into different subsets based on their availability, bioactivity, origin, and reactivity (Irwin et al. 2012).

2.3.2 Proprietary Databases

The leading pharmaceutical companies own large libraries of chemical substances along with a wide range of data including physicochemical properties and toxicity. This big data maintained by pharmaceutical companies can be the biggest source of hit identification for different therapeutic targets and information retrieved will help to optimize the lead compounds in the process of drug discovery and development. Recently, European Lead Factory, a consortium formed by the Innovative Medicines

Table 2.1 Chemical space navigation of compounds

Database	Company	Size	Chemical space
GSKchemspace XXL	GSK	10^{26}	Proprietary
MASSIV 2018	Merck	10^{20}	Proprietary
JFS	Johnson & Johnson	10^{19}	Proprietary
AstraZeneca	AstraZeneca	10^{17}	Proprietary
EvoSpace2016	Evotec	10^{16}	Proprietary
PGVL 2008	Pfizer	10^{14}	Proprietary
KnowledgeSpace	BiosolveIT	10^{14}	Public
BICLAIM 2012	Böehringer Ingelheim	10^{11}	Proprietary
GDB-17	University of Bern	10^{11}	Public
LPC 2016	Lilly	10^{10}	Proprietary
REAL Space	Enamine	10^{10}	Proprietary
Gala XI	WuXi	10^{9}	Commercial
ZINC 20	ZINC	10^{9}	Commercial
NCI SAVI	National Cancer Institute	10^{8}	Public
CHIPMUNK (CHemically feasible In silico Public Molecular UNiverse Knowledge base)	TU Dortmund University	10^{8}	Public
Pubchem	NCBI/NIH	10^{8}	Public
REAL Database	Enamine	10^{8}	Commercial
Aldrich Market	Sigma Aldrich	10^{8}	Commercial
Ultimate	MCule	10^{8}	Commercial
SCUBIDOO (Screenable Chemical Universe Based on Intuitive Data OrganizatiOn)	University of Marburg	10^{7}	Public
eMolecules Plus	eMolecules	10^{7}	Commercial

Initiative (IMI), is a joint initiative with the members of the European Federation of Pharmaceutical Industries and Associations (EFPIA) that provides access to approximately 550,000 novel compounds for screening molecular targets. (Karawajczyk et al. 2016). Among the different databases available (Table 2.1), the proprietary databases from pharmaceutical companies contain a larger number of compounds, and GSKchemspace XXL is one of the largest databases containing 10^{26} compounds.

2.3.3 Commercial Chemical Space

Due to restrictions in accessing proprietary databases to other researchers, in recent years several commercial databases have emerged to provide accessibility for screening libraries of molecules against the desired target in the drug discovery process. The virtual chemical spaces which provide billions of molecules are now under the reach of every medicinal chemist. Recent advancements in the computational-aided drug design, such as pharmacophore modeling and feature- and shape-based similarity search, help medicinal chemists for scaffold-hopping, hit expansion, and SAR exploration in intellectual property-free chemical space with minimum cost (BIG Data Center Members 2018).

One such example of a commercial database is Enamine which was established to meet the rising demand of high-throughput screening libraries. It has expertise in the synthesis of building blocks along with libraries of derived small molecular weight compounds for drug discovery needs. At present, it has over 225,000 original and unique compounds in stock, 210 million make-on-demand building blocks, and a virtual library of 14 billion REAL compounds (Enamine REAL Space and REAL Database [Available from: https://enamine.net/compound-collections/real-compounds/real-database]). Enamine also offers custom synthesis, compound library design, and lead optimization support to researchers in different stages of drug discovery. Another such example is MCule which enables researchers to find, optimize, and procure hit and lead compounds. The recently released ULTIMATE database consists of 122 million compounds and provides a comprehensive, curated library of molecules available for virtual screening (Kiss et al. 2012).

2.4 Data Analysis Tools

2.4.1 Artificial Intelligence

Artificial intelligence (AI) is a field of computer science that aims to develop efficient algorithms and can creatively solve problems that normally require human intelligence. After the successful application of AI in fields that includes image and speech recognition, traffic route suggestion, and weather forecasting, it has become a more interesting application in drug discovery and development. There are several stages of drug discovery such as target identification, hit identification, lead optimization, pharmacokinetic prediction, and clinical trial design where the application of AI becomes more advantageous (Fig. 2.3). Several computational models have been generated for activity and property prediction (QSAR or QSPR) with the aid of AI (Yang et al. 2019). The combination of AI with de novo drug design suggests different compounds datasets with synthetic feasibility and focused bioactivity (Lake 2019). Besides, AI provides great support for the progress of better candidate compounds into clinical trials, efficiently validate protein targets, and update clinical trial models to speed up the drug discovery program (Bender and Cortés-Ciriano 2020).

2.4.1.1 Machine Learning

Machine learning (ML) is a subset of AI with the capability to learn itself from the given data. In the last decade, there has been an exponential growth of machine learning tools such as Naïve Bayesian, support vector machines, pattern analysis, neural networks, and logistic regression (Bishop 2013). The machine learning models leverage the features of big data and allow the prediction of biological

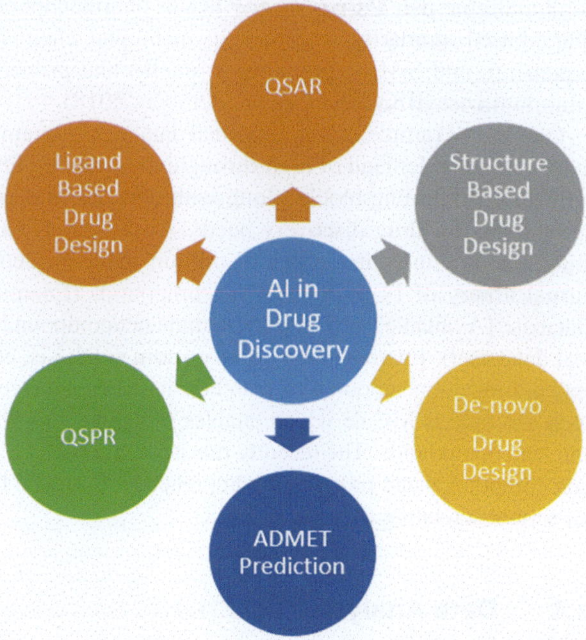

Fig. 2.3 AI in drug discovery. In drug discovery, AI is applied for structure-, ligand-, or de-novo-based drug design; QSAR OR QSPR studies; and ADMET prediction studies

activities of compounds and molecular properties with higher accuracy (Hessler and Baringhaus 2018). Machine learning methods have some limitations like the requirement of big labeled data for training the models, the data sparsity, and the limited scope of their interpretability which demands newer tools to address these problems (Elbadawi et al. 2020).

2.4.1.2 Deep Learning

Deep learning (DL) is a further subset of machine learning that learns to perform tasks with greater accuracy and minimal human involvement. It uses an artificial neural network in several layers to learn from data. Owing to the availability of big data from public and commercial databases, deep learning is gaining popularity in the drug discovery and development domain as well (Chen et al. 2018b). Deep learning differentiates from machine learning by the flexibility of architecture of their neural network. Several architectures of Deep Neural Networks (DNN) are in use such as a fully connected deep neural network that has multiple hidden layers in it, a convolutional neural network that contains several convolution layers and subsampling layers, recurrent neural network, and autoencoders which are used for unsupervised learning (Chen et al. 2018b). Several applications of DL models have been reported in property/activity prediction, reaction prediction, and retrosynthetic analysis, de novo design of ligands, and protein-ligand interaction prediction (Ekins et al. 2019).

2.4.2 Example Using AI for Data Analysis in the Pharmaceutical Company

2.4.2.1 BenevolentAI

Several pharmaceutical companies have started to apply AI in drug discovery and development for a variety of diseases in every step from hypothesis generation to early-stage clinical development. Among them, BenevolentAI is one particular mention that applies AI in drug discovery and development in different therapeutic areas alone and in collaboration with leading pharmaceutical companies. They use their proprietary platform in drug discovery projects which extract and analyze biomedical information using AI algorithms and create a knowledge-based graph. This automated platform hypothesizes and validates new ways to treat disease and personalize drugs, i.e., create precision medicine for patients (BenevolentAI [Available from: https://www.benevolent.com/]).

BenevolentAI's knowledge graphs are composed of structured medical information with over billions of relationships between genes, targets, diseases, protein, and drugs extracted from the scientific literature. Recently, it executed a drug repurposing program for searching antiviral drugs against COVID-19 and identified baricitinib as a potential candidate to decrease the ability of the virus to infect lung cells. The drug baricitinib, a rheumatoid arthritis drug by Eli Lilly, is now authorized by FDA for emergency application to patients who require supplement oxygen or invasive mechanical ventilation (Richardson et al. 2020).

2.4.2.2 Atomwise

Another example of a company that uses AI for structure-based small molecule drug discovery is Atomwise. It applies their proprietary AI platform, AtomNet, in drug discovery projects which use deep learning convolution neural network (CNN) for structure-based bioactivity prediction. It applies a statistical-based approach that extracts information from experimental affinity measurements of small molecules and proteins for predicting the binding affinity of small molecules to proteins. The AI approach implemented in AtomNet allows us to learn multidimensional data and extract the features which possibly classifies favorable and unfavorable associations that occurred between small molecules and proteins (Wallach et al. 2015). Atomwise partners in projects targeting different proteins for hit identification, hit-to-lead, and lead optimization and understanding the mechanism of action. The AtomNet technology by Atomwise has been proposed to unlock various undruggable targets till date and aim to do more in the future (Stecula et al. 2020).

2.4.3 Programming and Scripting Tools for Data Analysis

The integration of big data and the latest technology has allowed researchers to transform the time-consuming and cost-intensive traditional drug discovery programs into extremely efficient and cost-effective with the support of big data analytic tools and computational approaches. This advancement has been possible

Table 2.2 Data analytic tools used in big data analysis for drug discovery and development

Type	Tools	Website (accessed on February 1, 2021)
Analytics	SAS R Python	https://www.sas.com https://www.r-project.org https://www.python.org
Big data	Hadoop Hive MongoDB Cassandra	https://hadoop.apache.org https://hive.apache.org https://www.mongodb.com https://cassandra.apache.org
Support data analytic tools	Scala Spark MySQL	https://www.scala-lang.org https://spark.apache.org https://www.mysql.com
Data transformation	Informatica Abinitio	https://www.informatica.com https://www.abinitio.com
Visualization tools	Tableau TIBCO Spotfire PowerBI Matplotlib ggplot	https://www.tableau.com https://www.tibco.com https://powerbi.microsoft.com https://matplotlib.org https://ggplot2.tidyverse.org
Integrated development environment	PyCharm Jupyter Notebook, RStudio, Intellij	https://www.jetbrains.com/pycharm/ https://jupyter.org https://rstudio.com https://www.jetbrains.com
Web scraping	Beautiful Soup Scrapy	https://www.crummy.com/software/BeautifulSoup, https://scrapy.org
Cloud computing platforms	Amazon Web Services Microsoft Azure	https://aws.amazon.com, https://azure.microsoft.com
Business intelligence tools	Yellowfin BI MicroStrategy SAP Business Intelligence	http://www.yellowfinbi.com, http://www.microstrategy.com, http://www.sap.com
ETL tools	Hevo Data Luigi Talend	https://hevodata.com https://luigi-project.io https://www.talend.com
Data cleaning tools	Drake Trifacta Wrangler OpenRefine	https://www.drake.edu https://www.trifacta.com https://openrefine.org
Data collection tools	Forms on fire GoSpotCheck	https://www.formsonfire.com https://www.gospotcheck.com
Data modelling tools	Amundsen Lucidchart	https://www.amundsen.io https://www.lucidchart.com
Data migration tools	IRI NextForm Xplenty	https://www.iri.com https://www.xplenty.com

only due to the existence of different programming languages, analysis tools, computational platforms, etc. (Team RDC 2010). Several useful tools are summarized in Table 2.2, and some of the most used techniques are summarized in the following section.

Table 2.3 Python libraries useful for drug discovery and development

Library	Resource	Reference
RdKit	rdkit.org	Landrum (2016)
PyDPI	pypi.org/project/pydpi	Cao et al. (2013a)
PaDEL	pypi.org/project/padelpy	
HTSeq	htseq.readthedocs.io/en/master	Anders et al. (2015)
ODDT	pythonhosted.org/oddt	Wójcikowski et al. (2015)
AutoClickChem	autoclickchem.ucsd.edu	Durrant and McCammon (2012)
PyBioMed	projects.scbdd.com/pybiomed.html	Dong et al. (2018)
ChemoPy	code.google.com/archive/p/pychem/downloads	Cao et al. (2013b)
Scopy	github.com/kotori-y/Scopy	Yang et al. (2020)

2.4.3.1 Python

Python is an open-source, high-level general-purpose programming language that has become very popular in data science. Its popularity is because of its ease of use, flexibility, and availability of a large number of libraries for almost every field of data science (Stančin and Jović 2019). Python plays a pivotal role in automating the process of extracting and curating data from different resources such as chemical and biological databases, scientific literature published and thus aiding in the process of drug discovery and development. The numerical calculation of big data is executed with popular libraries such as NumPy (Oliphant 2006) and SciPy (Jones et al. 2001) and data generated using the library, Pandas (McKinney 2011) make Python as a programming language easy to learn and use for different applications in drug discovery and development. In addition, several libraries are available for data visualization, which include matplotlib (Hunter 2007), Plotly (Plotly Technologies Inc. 2021), seaborn (Anon. n.d), Bokeh (Team BD 2018), and ggplot2 (Wickham 2016). Some of the highly popular machine learning and deep learning libraries often applied for solving problems and generating models in drug discovery are scikit-learn (Pedregosa et al. 2011), TensorFlow (Abadi et al. 2016), Keras (Keras 2015), and PyTorch (Paszke et al. 2019). With growing developments of packages, Python is becoming popular in the area of drug discovery and development, and a few of the interesting packages are listed in Table 2.3.

2.4.3.2 R Package

R is another interesting and open-source language for statistical computing. Using this programming language a lot of cheminformatics tools are developed for accessing the big data which includes physicochemical properties, toxicity and bioactivity prediction, and clinical trial projects (Scot and Max 2012). Few examples to mention the application of R in drug discovery projects are the application of R-shiny in the QA process of Roche Diagnostics, rcdk, and cluster packages in the application of selecting candidate drugs for clinical trials (Voicu et al. 2020).

Table 2.4 List of R packages used in Clinical Trial Design available in cran-project.org

Package	Description
Atable	Create tables for reporting clinical trials
compareODM	Comparison of medical forms in CDISC ODM format
CRTSize	Sample size estimation in cluster (group) randomized trials
Blockrand	Creates randomizations for block random clinical trials
DoseFinding	Supports design and analysis of dose-finding experiments
Pact	Predictive analysis of clinical trials
SASxport	Read and write "SAS" "XPORT" files
ADCT	Adaptive Design in Clinical Trials
ClinPK, cpk	Clinical Pharmacokinetics Toolkit
randomizeR	Randomization for clinical trials
Base R	The functionalities are useful for the design and analysis of clinical trials
Greport	Graphical reporting for clinical trials

Table 2.4 provides the popular and important R packages used in clinical trial design.

2.4.3.3 SAS

Statistical Analysis System (SAS) is used widely for big data analysis in pharmaceutical industries due to its flexibility, ease to handle, and availability of a large number of automated pipelines for sorting, visualizing, and statistical analysis of data. Similar to python and R, SAS can be applied for predicting the bioactivity of molecules, decision-making in progressing drug candidates for a clinical trial, designing a clinical trial, and enhancing synergy and collaboration between the teams of drug discovery and development (Moser et al. 1988).

2.4.4 Example Using Programming Tools for Data Analysis

2.4.4.1 Exscientia

Exscientia is one of the emerging companies which utilizes the power of big data and AI for developing new drugs. Recently, it has been reported to identify a candidate drug, DSP-1181, for the treatment of obsessive compulsive disorder (OCD) in collaboration with a Japanese pharmaceutical company, Sumitomo Dainippon Pharma, using their proprietary AI platforms and big data. The molecule DSP-1181, a long-acting 5HT1A agonist, is currently under Phase I clinical trials and is more efficient in comparison to the endogenous agonist serotonin and current drugs available for OCD. The application of AI for the identification of DSP-1181 is considered a great success due to its short duration of discovery time, approximately 12 months in comparison to the traditional drug discovery approach which requires more than 10 years (Burki 2020).

2.4.4.2 AstraZeneca

AstraZeneca is another example of a pharmaceutical company that applies big data and AI in the drug discovery and development process. Through the application of AI, it finds new biological insights for various diseases by deciphering the wealth of information and predicting potential drug candidates, to score the images obtained from tumor and immune cells. Among different programming tools utilized for the development of computational models, PyDrone is one example that utilizes Python programming as back-end code for the molecular property calculator called Drone. The application of python helps in developing a quick and easy tool with a user-friendly environment that can adapt to a variety of prediction methods (Cao et al. 2013a). Currently, it collaborates with AI and molecular modeling companies including Mila, BenevolentAI, AI Sweden, Schrödinger, and consortia such as Machine Learning for Pharmaceutical Discovery and Synthesis (MLPDS) and Machine Learning Ledger Orchestration for Drug Discovery (MELLODDY) Consortium for pushing the limits of traditional drug discovery and development using advanced computational approaches. Among them, one example is the collaboration between AstraZeneca and BenevolentAI focused to enhance the use of AI and ML in drug discovery and development for chronic kidney disease and idiopathic pulmonary fibrosis by combining the strengths of AstraZeneca in genomics, chemistry, and clinical trials with BenevolentAI's target discovery platform and biomedical knowledge graph (AstraZeneca Starts Artificial Intelligence Collaboration To Accelerate Drug Discovery [Available at https://www.benevolent.com/news/astrazeneca-starts-artificial-intelligence-collaboration-to-accelerate-drug-discovery]).

2.5 Drug Screening Platform

Drug discovery and development is a complicated process that is often tedious and time-consuming. The entire process is highly uncertain and can hit a roadblock at any step: right from the early stages, i.e., hit identification, to later stages, i.e., clinical trials of therapeutic candidates (Fig. 2.4). The biggest challenge in drug development is hit identification (Hughes et al. 2011). These challenges can be addressed by the introduction of various high-throughput technologies in the drug screening platform. In recent decades, drug screening platforms based on big data are being used in various drug discovery programs. It has helped to analyze a large volume of diverse datasets of chemical compounds in order to better understand the disease and discover new chemical entities (Chen and Butte 2016). Nevertheless, developing an effective therapeutic treatment strategy for complex diseases has also been challenging because of the complicated pathophysiological nature of the disease. In order to overcome these challenges, web-based drug screening platform has been developed by integrating AI to the disease's pathological features and thus by facilitating the productive and economical drug discovery process (Park et al. 2021). The use of HTS robotics, along with the automation in data analysis and validation systems, is the characteristics of an ideal drug screening platform as the researcher could efficiently conduct multiple experiments to identify hits (https://sites.tufts.edu/INSciDE/drug-screening/). Some of the traditionally used efficient screening platforms are 2D cellular/tissue-based systems in micro-well plates, whereas modern technology focused on 3D

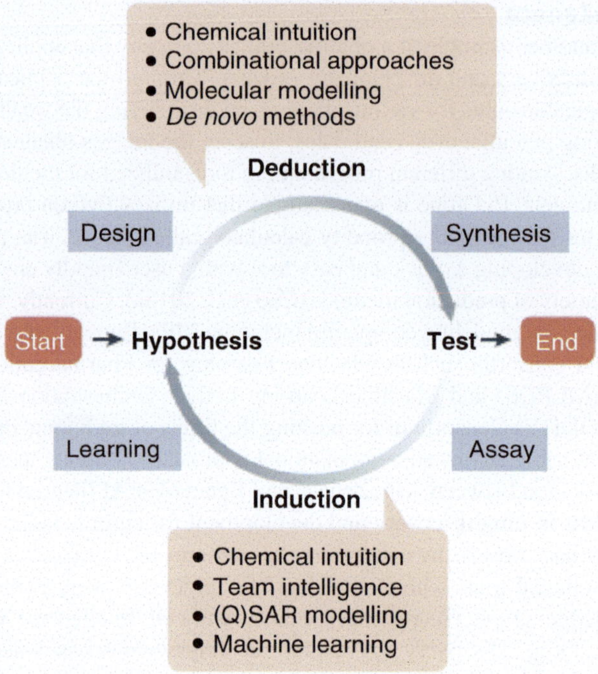

Fig. 2.4 The molecular design cycle. The molecular design cycle initiates with obtaining hit molecules in the HTS platform, and the discovery process eventually leads to the optimized preclinical candidate compounds as the final product. Smart automation of the various events in the molecular design cycle can help to save time as well as reduces error, thereby resulting in more productive and efficient drug discovery. For example, advanced lab-on-a-chip technology, coupled with machine learning methods, represents modern and efficient drug discovery technology. The whole design cycle can be controlled completely by programmed software (Schneider 2018)

microfluidics or even chip-based methods which drastically reduce the effort for analyzing large volume of samples as well as mimic the natural in vivo conditions.

2.5.1 Types of HTS Screening Platform

The two common types of screening platforms are phenotypic (whole-cell screening) and target-based biochemical screening. Phenotypic screening evaluates the activity of chemical compounds in vivo (in whole animals), or ex vivo (isolated tissues/organs), whereas target-based screening involves testing the molecules on purified target proteins (Kiriiri et al. 2020). The phenotypic screening method is used to identify molecules with certain pharmacological properties but doesn't elucidate the basic molecular mechanism of action. On the other hand, target-based screens are focused also on a molecular mechanism of action. The phenotypic drug screening method is more physiologically relevant as they provide accurate information about the therapeutic potential of the drug in an earlier stage of drug discovery as compared

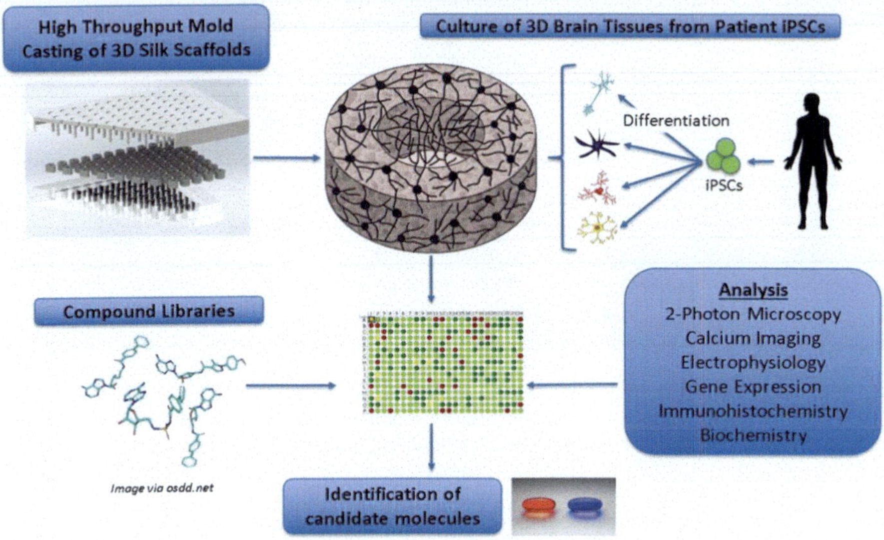

Fig. 2.5 Example of a drug screening platform for neurological disorders. This screening platform is developed by the Initiative for Neural Science, Disease & Engineering (INSciDE@Tufts) at Tufts University (https://sites.tufts.edu/INSciDE/drug-screening/). This approach has wider applicability as it can be used for the various tissue-based platform

to target-based screening which is more rational, requiring isolation of the target (Kiriiri et al. 2020).

The HTS platform system shown in Fig. 2.5 is based on 3D bioengineered brain tissue models. The platform mimics native in vivo conditions and has also high accuracy of hits identification. One of the biggest advantages of this platform is that these 3D bioengineered tissues can be maintained in culture for more than a year and the systems are integrated for both real-time imaging (i.e., data measurement) and terminal data analysis methods. The platform can image changes in metabolism and tissue structure, as well as can measure neuronal signals in order to analyze metabolites. Overall, this platform is highly useful for the high-throughput in vitro study of neurological disorders (https://sites.tufts.edu/INSciDE/drug-screening/).

2.5.2 Screening Platform in Pharmaceutical Industries

HTS screening platforms in the pharmaceutical/biotech industry should cover large areas of chemical diversity. In pharmaceutical industries, these compounds are tested against different biological targets usually in 1536- or 3456-well plates. The most efficient and frequently used assay system in the HTS platform is colorimetric, luminescence, fluorescence, immunoassays, and radioligand binding assays. These methods facilitate simple and convenient assay procedures and have high robustness, sensitivity, and specificity (Szymański et al. 2012). In big industries, the whole

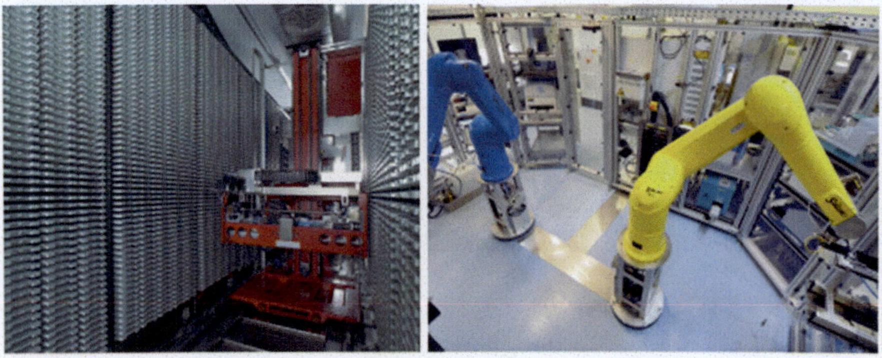

Fig. 2.6 Automated drug screening platform facilities. Boehringer–Ingelheim Pharma has the capacity to store millions of chemical compounds in compact facilities and is handled by robots for high-throughput assays against the biological target of interest (Schneider 2018)

process is automated and integrated with robotics because of which there is a capacity to screen thousands of compounds in a single day (Fig. 2.6). There is also a possibility to use a wide variety of HTS bioassays capable of measuring various cellular functions as well as biochemical functions. HTS of pharmaceutical compound libraries continues to dictate the discovery of new chemical entities and was the most important drug discovery method for numerous commercial drugs, e.g., sitagliptin (Januvia), an antidiabetic medication which is the dipeptidyl peptidase-4 inhibitor, and rivaroxaban (Xarelto), blood thinner agent which is a clotting factor XA inhibitor (Szymański et al. 2012). The important characteristics of pharmaceutical drug screening platforms are the size or the number of chemical compounds in the compound screening deck, purity, and ease of chemical synthesis, physicochemical characteristics of compounds including drug-likeness or lead-likeness, and chemical diversity to uncover numerous chemical spaces (Szymański et al. 2012). Generating a sufficient novel scaffold for patenting the compound is a phenomenal task in the pharmaceutical industry. There can be a considerable challenge on patenting while working on the old databases of the HTS collection or while working with the proprietary database of compounds (e.g., zinc, enamine, and Cambridge compound databases). The increase in the size of a screening deck with the compounds and drug-like properties can strengthen the HIT identification process (Szymański et al. 2012). Chemokine receptor antagonist, maraviroc (Selzentry for USA / Celsentri for EU developed by Pfizer), is an example where the hits were discovered by screening an in-house compound library of molecules. In 1997, Pfizer initiated the drug discovery program of HTS with an in-house library of about 500,000 compounds using a radioligand binding assay developed for CC-chemokine receptor 5 (CCR5) and identified a weak agonist hit with weak cellular antiviral activity. However, this hit molecule aided as a lead molecule and is further optimized by synthesizing over 4000 compounds, analyzing their SAR, and improving PK and PD properties including removal of undesired cytochrome and potassium channel activity with slow offset antagonist activity. The final optimized compound, maraviroc, was identified as the suitable clinical candidate with high antiviral potency and prolonged receptor occupancy and was approved by FDA in 2007 (Macarron et al. 2011). At present, the HTS screening deck of the biggest pharmaceutical companies comprises millions of compounds and further need of enhancement programs for increasing the number of compounds might be limited due to the investment constraints. On the other hand, the investment should not be aimed only at a numerical increase of the compound deck but should also be aimed at including new prototypes of chemical compounds which are bound to be effective against various current undruggable targets, for example, novel enzymes target, protein-protein interactions target, gap junction protein, and various transporters. The diversity in the compound deck may be introduced by including chemotypes that are different from those of the past, for example, natural product mimics, macrocyclic biomolecules, and conjugate drugs. (Macarron et al. 2011).

2.5.3 Commercially Available Screening Platform

HTS screening platforms are highly effective in identifying lead compounds for drug development, but high costs of hosting, maintaining, and distributing chemical libraries are the limiting factors. Nonetheless, for the small- and medium-sized drug discovery efforts for which the maintenance of an in-house compound deck is expensive, the commercially available screening platform provides a good alternative. Ubiquigent (Ubiquigent-Drug discovery screening platform, https://www.ubiquigent.com/drug-discovery-screening-platform/) is a commercial and advanced drug screening platform for studying ubiquitylation and protein degradation in drug discovery. This platform provides HTS assays for identifying deubiquitylase (DUB) enzyme inhibitors, modulators of ubiquitin ligases, and modulators of the ubiquitin-proteasome system (UPS) degradation via (proteolysis targeting chimera) PROTAC. They provide PROTEOMEprofiler™ and DUBprofiler-Cell™ platforms which are coupled with tandem mass spectrometry-enabled platforms to support target identification and validation. Both the platforms can rapidly reveal HIT molecules and provide support to QSAR of compounds leading to optimization of potency and selectivity across all DUBs (Ubiquigent-Drug discovery screening platform, https://www.ubiquigent.com/drug-discovery-screening-platform/).

2.6 Data Analysis in Drug Screening Platform

The data obtained from the HTS assays can be stored on the public as well as on proprietary databases because of which it is possible to revisit and retrieve the data. Scientific and systematic data management of drug screening platforms can be used to train machine learning on various information like ligand–target interaction, target selectivity, ligand–pharmacological effect, and ligand physiochemical properties-pharmacological effect. The machine learning approach helps to build innovative and robust multidimensional QSAR which can even suggest chemical

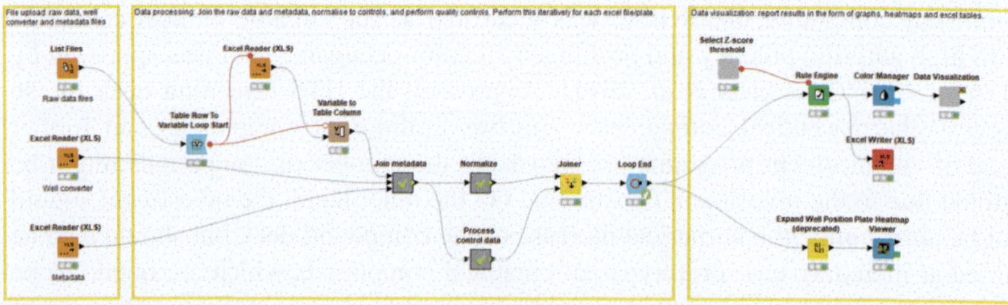

Fig. 2.7 Overview of the high-throughput screening workflow in KNIME HUB. The first step in the hub is file uploading where raw data, well converter, and metadata files are uploaded to the hub. The second step involves data processing where the raw and metadata are joined, then control is normalized, and quality controls checks are performed. The third step involves data visualization resulting from graphs, heat maps, or tables (Berthold et al. 2009; Hansel et al. 2020)

compounds for further testing (Schneider 2018). The algorithms in machine learning help to recognize hidden properties in chemical compounds that aren't typically recognized because of the large set of compounds and their viability (Schneider 2018). The most advanced drug screening platform based on HTS screening can even suggest suitable chemical structures with inputs for chemical synthesis. One of the most advanced open-source machine learning software developed by Google is "TensorFlow" (Abadi et al. 2016) (https://www.tensorflow.org/) which has found widespread use and advantage for drug discovery. The deep learning system of HTS databases provides meaningful structure–activity relationships between ligands, targets, and genotypes and eventually provides an accurate prediction for drug repurposing, drug development, and target identification (Schneider 2018). A new comprehensive database of chemogenomics data analysis has been recently published by a collaboration of industrial and academic partners (ExCAPE-DB212, Exascale Compound Activity Prediction Engine) (Sun et al. 2020). KNIME Analytics Platform (Berthold et al. 2009) is one of the analysis and visualization platforms that is designed to handle large volume of HTS data conducted on very large libraries of the chemical compounds. This workflow automatically imports HTS data and processes it to identify HIT molecules with desirable qualities, and it has also the robustness to identify false positive and negative results. The main advantage of this kind of hub is to enable rapid visualization and interpretation of the screening results (Fig. 2.7, workflow can be download at Hansel et al. 2020) (Bender and Cortés-Ciriano 2020).

2.6.1 Case Studies of HTS Drug Screening Platform

2.6.1.1 Case Study 1: ***The Deconstructed Granuloma: A Complex High-Throughput Drug Screening Platform for the Discovery of Host-Directed Therapeutics against Tuberculosis***

In this study, Huang et al. describe a new screening platform called "deconstructed granuloma" based on murine TB granulomas which use fluorescent *Mycobacterium tuberculosis* reporter strains to screen the infected host immune environment. In this study, the authors used known antibacterial compounds in 384-well plates in order to validate the screening platform. The authors specified that this HTS platform is effective, sensitive, selective, and highly robust for screening compounds against various immunological targets (Huang et al. 2018).

2.6.1.2 Case Study 2: ***Drug-Screening Platform Based on the Contractility of Tissue-Engineered Muscle***

In this study, Vandenburgh et al. developed a tissue-based HTS platform to screen chemical compounds in vitro system. In this study, skeletal muscles in the myoblasts was engineered into a 3D system with parallel myofibers. The growth of the tissue in a 96-well plate generated tetanic forces upon electrical stimulation which are then measured with a motion detection system. The tetanic active forces are increased in response to insulin-like growth factor 1 and are deteriorated with a cholesterol-lowering statin. Overall, this study integrates tissue engineering and biomechanical measurement of a chemical compound into a single platform for screening (Vandenburgh et al. 2008).

2.6.1.3 Case Study 3: ***Development of a Drug Screening Platform Based on Engineered Heart Tissue***

In this study, Hansen et al. developed a drug screening platform using engineered heart tissue (EHT) and evaluate the contractile activity (Hansen et al. 2010). This platform used neonatal rat heart cells which are then mixed with fibrinogen and thrombin and then pipetted into rectangular casting molds. Thus, developed contractile activity is monitored video-optically by a camera and evaluated. Fibrin-based mini-EHTs (FBMEs) used in these experiments were then transferred to a standard 24-well plate. Over time FBMEs condensed to form a gel, and after 8–10 days, FBMEs started to rhythmically deflect the posts which allowed calculation of rate, force, and kinetics. The authors optimized the study with a large series of FBME ($n = 192$), with the results indicating an increase in relaxation time with chromanol, quinidine, and erythromycin decrease with doxorubicin.

2.6.1.4 Case Study 4: ***Developing a Drug Screening Platform: MALDI-Mass Spectrometry Imaging of Paper-Based Cultures***

Tobias et al. developed paper-based cultures (PBCs) as a 3D platform in which cells suspended in Matrigel are seeded into paper scaffolds in order to replicate a tissue-like environment (Tobias et al. 2019). In this study, the authors demonstrate the potential of matrix-assisted laser desorption/ionization–mass spectrometry imaging

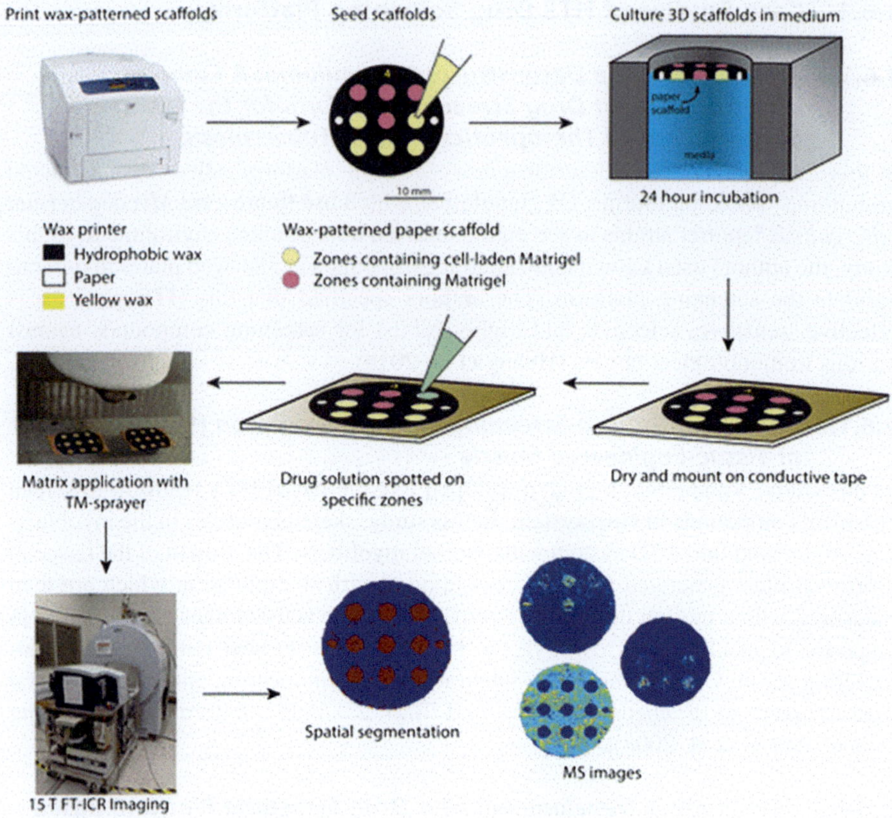

Fig. 2.8 Workflow of matrix-assisted laser desorption/ionization–mass spectrometry imaging with paper-based cultures (MALDI-MSI-PBC) as a drug screening platform. In this platform, wax patterns are printed on Whatman paper, then patterned scaffolds are seeded with Matrigel followed by 24 h incubation in a culture medium. Then the patterned scaffolds are dried and mounted onto a MALDI. Next, the chemical compounds for testing in the form of solutions were applied, then allowed to redry, followed by spraying of Matrix. Lastly, the analysis was done using MALDI-MSI (Tobias et al. 2019). Copyright 2019 American Chemical Society

with PBCs (MALDI-MSI-PBC) as a drug screening platform. This method discriminated regions of the PBCs with and without drugs, elucidating the potential of this platform to develop rapid, large-scale, and parallel mass spectrometric drug screening methods (Fig. 2.8).

2.6.1.5 Case Study 5: *A Logical Network-Based Drug Screening Platform for Alzheimer's Disease Representing Pathological Features of Human Brain Organoids*

Alzheimer's disease (AD) has a complex pathology because of which the drug development effort is challenging. Park et al. developed an efficient, network-based drug screening platform integrating mathematical modeling with the

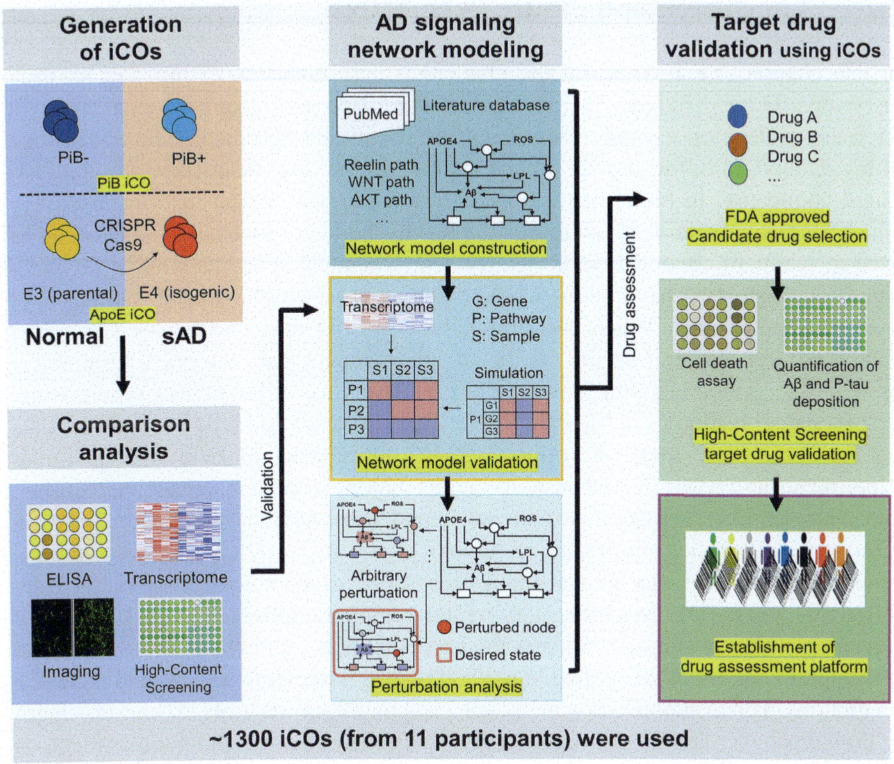

Fig. 2.9 A network-based drug screening platform for Alzheimer's disease. To establish drug screening platform, iPSC-derived organoids (iCOs) and CRISPR-Cas9 lines were used. The steps used in the development of the platform are systems biology-based AD pathway simulation, signaling network construction, network model validation, and identification of control nodes. The platform was validated by drugs using FDA-approved drugs (Park et al. 2021)

pathological features of AD (Fig. 2.9) (Park et al. 2021). Human iPSC-derived cerebral organoids (iCOs) and CRISPR-Cas9-edited isogenic lines were used to build the high-throughput screening platform. The platform was validated by studies involving FDA-approved drugs. The platform was also used to access the blood–brain barrier (BBB) permeability of chemical compounds since AD is a neurodegenerative disease and for its successful treatment sufficient level of the drug needs to be present in the brain by crossing BBB. Overall, their study presents a novel strategy for HTS drug screening platform for AD by the convergence of mathematical modeling and the pathological feature of AD.

2.7 Conclusion and Future Perspective

In this chapter, we focused on the challenges and advances of big data in drug discovery. Drug discovery is a complicated process and requires an average of 15 years to develop a single drug using the traditional approach. The application of big data could tremendously reduce the timeline in drug discovery. Big data has found application in both chemical as well as biological aspect of drug discovery which is clearly explained with few examples. In chemistry focus, the big data can be used to maintain chemical deck management, chemical space as well as physico-chemical properties navigation, QSAR studies, and combinatorial synthesis of chemical compounds. When concerning biology, big data can be used in target validation, hit screening, the study of pharmacological activities, genomics-proteomics, conducting clinical trials, PK-PD studies, as well as in drug repurposing. In this chapter resources of big data for the drug discovery are reported, and the data can be retrieved to make the drug discovery more efficient by avoiding similar experiments and cost-effectiveness. AI and its major components, machine learning, and deep learning are described in detail and illustrated its application in drug discovery. AI can help in predicting biological affinity, activity, and physiological properties based on the data available for series of chemical compounds from different repositories. Machine learning found its usefulness in high-throughput screening where it can also suggest the compounds for chemical screening. Deep learning models are often applied in predicting drug-target interactions and ADMET predictions. The available software, scripts, and data analysis tools for various stages of drug discovery are greatly needed and the currently existing and popular methods are reported. The major obstacle in drug discovery is hit identification, and this can be enhanced by the application of AI in drug screening platforms. Additionally, the HTS screening platform, big data management of the HTS screening platform, and data analysis of HTS screening are discussed in detail. The application of AI and robotics in screening platforms greatly improved the quality and quantity of data available in drug discovery. These approaches are presented using case studies where the use of AI resulted in a robust screening platform for efficient and accurate identification of hits. Overall an overview of big data application in various stages of drug discovery helps to understand the advancements in pharmaceutical and biotechnological research.

Currently, the application of big data is being applied only at limited stages of the drug discovery process; if it has been applied at every single step of drug discovery, then it can result in a great impact in the whole process. In the future, the use of big data in the drug discovery process will possibly reduce the current timeline of the drug discovery process. The application of big data might transform the traditional drug discovery approach to efficient and smart drug discovery. Scientific data storage accelerates the drug development processes; however, data management systems available to meet the increasing quantity of scientific information are questionable. Appropriately designed platforms to support continuously evolving and increasing research data are urgently needed to bring drastic changes in the drug discovery process. As we look forward to the future of drug discovery, better AI approaches integrated with novel analytical mining techniques hold the key to faster and successful drug discovery and development programs.

References

Abadi M, Barham P, Chen J (2016) et al. TensorFlow: a system for large-scale machine learning. Proceedings of the 12th USENIX conference on Operating Systems Design and Implementation; Savannah, GA, USA: USENIX Association. p. 265–83

Agrawal R, Prabakaran S (2020) Big data in digital healthcare: lessons learnt and recommendations for general practice. Heredity 124:525–534

Anders S, Pyl PT, Huber W (2015) HTSeq—a Python framework to work with high-throughput sequencing data. Bioinformatics 31:166–169

Anon. (n.d) seaborn: statistical data visualization 2021 [Available from: https://seaborn.pydata.org/

Archenaa J, Anita EAM (2015) A survey of big data analytics in healthcare and government. Procedia Comput Sci 50:408–413

Auletta AE, Brown M, Wassom JS et al (1991) Current status of the Gene-Tox Program. Environ Health Perspect 96:33–36

Bender A, Cortés-Ciriano I (2020) Artificial intelligence in drug discovery: what is realistic, what are illusions? Part 1: ways to make an impact, and why we are not there yet. Drug Discov Today 26:511

Berthold M, Cebron N, Dill F et al (2009) KNIME: the Konstanz information miner. First publ in: Data Analysis, Machine Learning and Applications: Proceedings of the 31st Annual Conference of the Gesellschaft für Klassifikation eV, Albert-Ludwigs-Universität Freiburg, March 7–9, 2007 New York: Springer, 2008 V

BIG Data Center Members (2018) Database resources of the BIG Data Center in 2018. Nucleic Acids Res 46:D14–D20

Bishop CM (2013) Model-based machine learning. Philos Trans A Math Phys Eng Sci 371: 20120222

Blake RA (2007) Target validation in drug discovery. Methods Mol Biol 356:367–377

Bosc N, Felix E, Arcila R et al (2020) MAIP: a prediction platform for predicting blood-stage malaria inhibitors. Res Square

Burki T (2020) A new paradigm for drug development. Lancet Digit Health 2:e226–e2e7

Cáceres EL, Tudor M, Cheng AC (2020) Deep learning approaches in predicting ADMET properties. Future Med Chem 12:1995–1999

Cao D-S, Liang Y-Z, Yan J et al (2013a) PyDPI: freely available Python Package for chemoinformatics, bioinformatics, and chemogenomics studies. J Chem Inf Model 53:3086–3096

Cao D-S, Xu Q-S, Hu Q-N et al (2013b) ChemoPy: freely available python package for computational biology and chemoinformatics. Bioinformatics 29:1092–1094

Chen B, Butte AJ (2016) Leveraging big data to transform target selection and drug discovery. Clin Pharmacol Ther 99:285–297

Chen H, Engkvist O, Wang Y et al (2018b) The rise of deep learning in drug discovery. Drug Discov Today 23:1241–1250

Chen J, Luo X, Qiu H et al (2018a) Drug discovery and drug marketing with the critical roles of modern administration. Am J Transl Res 10:4302–4312

Cui JJ, Tran-Dubé M, Shen H et al (2011) Structure based drug design of Crizotinib (PF-02341066), a potent and selective dual inhibitor of mesenchymal–epithelial transition factor (c-MET) kinase and anaplastic lymphoma kinase (ALK). J Med Chem 54:6342–6363

Dickson M, Gagnon JP (2004) Key factors in the rising cost of new drug discovery and development. Nat Rev Drug Discov 3:417–429

Dong J, Yao Z-J, Zhang L et al (2018) PyBioMed: a python library for various molecular representations of chemicals, proteins and DNAs and their interactions. J Cheminf 10:16
Dunbar JB Jr, Smith RD, Damm-Ganamet KL et al (2013) CSAR data set release 2012: ligands, affinities, complexes, and docking decoys. J Chem Inf Model 53:1842–1852
Durrant JD, McCammon JA (2012) AutoClickChem: click chemistry in silico. PLoS Comput Biol 8:e1002397
Ekins S, Puhl AC, Zorn KM et al (2019) Exploiting machine learning for end-to-end drug discovery and development. Nat Mater 18:435–441
Elbadawi M, Gaisford S, Basit AW (2020) Advanced machine-learning techniques in drug discovery. Drug Discov Today 26:769
Fonger GC (1995) Hazardous substances data bank (HSDB) as a source of environmental fate information on chemicals. Toxicology 103:137–145
Frantzi M, Latosinska A, Mischak H (2019) Proteomics in drug development: the Dawn of a new era? Proteomics Clin Appl 13:e1800087
Gaulton A, Bellis LJ, Bento AP et al (2012) ChEMBL: a large-scale bioactivity database for drug discovery. Nucleic Acids Res 40:D1100–D11D7
Gilson MK, Liu T, Baitaluk M et al (2016) BindingDB in 2015: a public database for medicinal chemistry, computational chemistry and systems pharmacology. Nucleic Acids Res 44:D1045–D1D53
Glicksberg BS, Li L, Chen R et al (2019) Leveraging big data to transform drug discovery. Methods Mol Biol (Clifton, NJ) 1939:91–118
Gombar VK, Silver IS, Zhao Z (2003) Role of ADME characteristics in drug discovery and their in silico evaluation: in silico screening of chemicals for their metabolic stability. Curr Top Med Chem 3:1205–1225
Hansel C, Yousefian S, Carreras-Puigvert J (2020) High-throughput screening, data analysis, processing, and hit identification. https://www.knime.com/blog/a-workflow-for-high-throughput-screening-data-analysis-processing-and-hit-identification
Hansen A et al (2010) Development of a drug screening platform based on engineered heart tissue. https://www.ahajournals.org/doi/10.1161/circresaha.109.211458
Hessler G, Baringhaus K-H (2018) Artificial intelligence in drug design. Molecules 23:2520
Hoofnagle JH, Serrano J, Knoben JE et al (2013) LiverTox: a website on drug-induced liver injury. Hepatology 57:873–874
Huang L, Kushner NL, Theriault ME et al (2018) The deconstructed granuloma: a complex high-throughput drug screening platform for the discovery of host-directed therapeutics against tuberculosis. Front Cell Infect Microbiol 8:275
Hughes JP, Rees S, Kalindjian SB et al (2011) Principles of early drug discovery. Br J Pharmacol 162:1239–1249
Hulsen T, Jamuar SS, Moody AR et al (2019) From big data to precision medicine. Front Med 6:34
Hunter J (2007) Matplotlib: a 2D graphics environment. Comput Sci Eng 9:90–95
Irwin JJ, Sterling T, Mysinger MM et al (2012) ZINC: a free tool to discover chemistry for biology. J Chem Inf Model 52:1757–1768
Jones E, Oliphant T Peterson P (2001) SciPy: Open source scientific tools for Python
Kar S, Leszczynski J (2020) Open access in silico tools to predict the ADMET profiling of drug candidates. Expert Opin Drug Discovery 15:1473–1487
Karawajczyk A, Orrling KM, de Vlieger JS et al (2016) The European Lead factory: a blueprint for public-private partnerships in early drug discovery. Front Med (Lausanne) 3:75
Keras [Internet]. GitHub. 2015. https://github.com/fchollet/keras
Kilkenny M, Robinson K (2018) Data quality: "Garbage in—garbage out". Health Inf Manag 47: 183335831877435
Kim S, Chen J, Cheng T et al (2020) PubChem in 2021: new data content and improved web interfaces. Nucleic Acids Res 49:D1388–D1D95
Kiriiri GK, Njogu PM, Mwangi AN (2020) Exploring different approaches to improve the success of drug discovery and development projects: a review. Future J Pharm Sci 6:27

Kiss R, Sandor M, Szalai FA (2012). http://Mcule.com: a public web service for drug discovery. J Cheminf 4:P17-P

Kitchin R, McArdle G (2016) What makes big data, big data? Exploring the ontological characteristics of 26 datasets. Big Data Soc 3:2053951716631130

Kopec KK, Bozyczko-Coyne D, Williams M (2005) Target identification and validation in drug discovery: the role of proteomics. Biochem Pharmacol 69:1133–1139

Lake F (2019) Artificial intelligence in drug discovery: what is new, and what is next? Future Drug Discov 1:FDD19

Landrum G (2016) RDKit: Open-Source Cheminformatics Software

Luo J, Wu M, Gopukumar D et al (2016) Big data application in biomedical research and health care: a literature review. Biomed Inf Insights 8:1–10

Lusher SJ, McGuire R, van Schaik RC et al (2014) Data-driven medicinal chemistry in the era of big data. Drug Discov Today 19:859–868

Macarron R, Banks MN, Bojanic D et al (2011) Impact of high-throughput screening in biomedical research. Nat Rev Drug Discov 10:188–195

McKinney W (2011) pandas: a foundational Python library for data analysis and statistics. Python High Performance Science Computer

Mendez D, Gaulton A, Bento AP et al (2018) ChEMBL: towards direct deposition of bioassay data. Nucleic Acids Res 47:D930–DD40

Mirza B, Wang W, Wang J et al (2019) Machine learning and integrative analysis of biomedical big data. Genes (Basel) 10

Mohs RC, Greig NH (2017) Drug discovery and development: role of basic biological research. Alzheimer's & dementia (New York, N Y) 3:651–657

Moingeon P (2021) Applications of artificial intelligence to new drug development. Ann Pharm Fr 79:566

Moser EB, Saxton AM, Geaghan JP (1988) Biological applications of the SAS system: an overview. Comput Appl Biosci 4:233–238

Oliphant T. Guide to NumPy 2006

Park JC, Jang SY, Lee D et al (2021) A logical network-based drug-screening platform for Alzheimer's disease representing pathological features of human brain organoids. Nat Commun 12:280

Paszke A, Gross S, Massa F et al. (2019) PyTorch: an imperative style, High-Performance Deep Learning Library

Paul SM, Mytelka DS, Dunwiddie CT et al (2010) How to improve R&D productivity: the pharmaceutical industry's grand challenge. Nat Rev Drug Discov 9:203–214

Pedregosa F, Varoquaux G, Gramfort A et al (2011) Scikit-learn: machine learning in Python. J Mach Learn Res 12:2825–2830

Plotly Technologies Inc. (2021). https://plot.ly

Poon CC, Kelly JJ (2017) Development of crizotinib, a rationally designed tyrosine kinase inhibitor for non-small cell lung cancer. Int J Cancer 140:1945–1954

Reimer AP, Madigan EA (2018) Veracity in big data: how good is good enough. Health Informatics J 25:1290–1298

Richardson P, Griffin I, Tucker C et al (2020) Baricitinib as potential treatment for 2019-nCoV acute respiratory disease. Lancet 395:e30–ee1

Ristevski B, Chen M (2018) Big data analytics in medicine and healthcare. J Integr Bioinform 15: 20170030

Roth BL, Lopez E, Patel S et al (2000) The multiplicity of serotonin receptors: uselessly diverse molecules or an embarrassment of riches? Neuroscientist 6:252–262

Russell C, Rahman A, Mohammed AR (2013) Application of genomics, proteomics and metabolomics in drug discovery, development and clinic. Ther Deliv 4:395–413

Schneider G (2018) Automating drug discovery. Nat Rev Drug Discov 17:97–113

Scot M, Max K (2012) The use of the R language for medicinal chemistry applications. Curr Trends Med Chem 12:1957–1964

Shaw AT, Solomon B, Kenudson MM (2011) Crizotinib and testing for ALK. J Natl Compr Cancer Netw 9:1335–1341

Stančin I, Jović A (2019) Editors. an overview and comparison of free python libraries for data mining and big data analysis. 2019 42nd International Convention on Information and Communication Technology. Electron Microelectron (MIPRO) 2019:20–24

Stecula A, Hussain MS, Viola RE (2020) Discovery of novel inhibitors of a critical brain enzyme using a homology model and a deep convolutional neural network. J Med Chem 63:8867–8875

Sterling T, Irwin JJ (2015) ZINC 15—ligand discovery for everyone. J Chem Inf Model 55:2324–2337

Sun J et al (2020) ExCAPE-DB: an integrated large scale dataset facilitating Big Data analysis in chemogenomics. https://jcheminf.biomedcentral.com/articles/10.1186/s13321-017-0203-5

Szymański P, Markowicz M, Mikiciuk-Olasik E (2012) Adaptation of high-throughput screening in drug discovery-toxicological screening tests. Int J Mol Sci 13:427–452

Talevi A, Bellera CL (2020) Challenges and opportunities with drug repurposing: finding strategies to find alternative uses of therapeutics. Expert Opin Drug Discovery 15:397–401

Team BD (2018) Bokeh: Python library for interactive visualization

Team RDC (2010) A language and environment for statistical computing: reference index. R Foundation for Statistical Computing, Vienna, p 2010

Tobias F, McIntosh JC, LaBonia GJ et al (2019) Developing a drug screening platform: MALDI-mass spectrometry imaging of paper-based cultures. Anal Chem 91:15370–15376

Vandenburgh H et al (2008) Drug-screening platform based on the contractility of tissue-engineered muscle. https://onlinelibrary.wiley.com/doi/10.1002/mus.20931

Voicu A, Duteanu N, Voicu M et al (2020) The rcdk and cluster R packages applied to drug candidate selection. J Cheminf 12:3

Vuong HG, Nguyen TQ, Nguyen HC et al (2020) Efficacy and safety of Crizotinib in the treatment of advanced non-small-cell lung cancer with ROS1 rearrangement or MET alteration: a systematic review and meta-analysis. Target Oncol 15:589–598

Wallach I, Dzamba M, Heifets A (2015) AtomNet: a deep convolutional neural network for bioactivity prediction in structure-based drug discovery. arXiv:151002855. [cs, q-bio, stat]

Wang Y, Bryant SH, Cheng T et al (2017) PubChem BioAssay: 2017 update. Nucleic Acids Res 45: D955–Dd63

Wickham H (2016) ggplot2: elegant graphics for data analysis. Springer-Verlag, New York

Wójcikowski M, Zielenkiewicz P, Siedlecki P (2015) Open drug discovery toolkit (ODDT): a new open-source player in the drug discovery field. J Cheminf 7:26

Yang X, Wang Y, Byrne R et al (2019) Concepts of artificial intelligence for computer-assisted drug discovery. Chem Rev 119:10520–10594

Yang Z-Y, Yang Z-J, Lu A-P et al (2020) Scopy: an integrated negative design python library for desirable HTS/VS database design. Briefings in Bioinformatics

Zang Q, Mansouri K, Williams AJ et al (2017) In silico prediction of physicochemical properties of environmental chemicals using molecular fingerprints and machine learning. J Chem Inf Model 57:36–49

Zhu H (2020) Big data and artificial intelligence modeling for drug discovery. Annu Rev Pharmacol Toxicol 60:573–589

An Overview of Databases and Tools for lncRNA Genomics Advancing Precision Medicine

3

Amarinder Singh Thind, Karambir Kaur, and Isha Monga

Abstract

The completion of the human genome project revealed many elusive questions about the "junk DNA" in the genome. One of the promising observations was the pervasive transcription and discovery of noncoding RNAs, which put aside the assumption of RNA as a mere messenger in the central dogma of molecular biology. It became evident that RNA plays more critical regulatory roles than simply being the genetic information carrier. The advancements in the next-generation sequencing methods in the last decade identified large numbers of long noncoding RNAs (lncRNAs) expressed during different conditions and performing diverse functions. Despite the rapidly growing advancements in various omics techniques to identify, annotate, and analyze the role of lncRNAs during different diseases, their molecular functions are still unexplored. To keep the pace of rapidly growing data and utilize them in novel lncRNA prediction, numerous computational repositories and predictive algorithms have been published in order to expand our horizons of the understanding and functions

Amarinder Singh Thind, Karambir Kaur and Isha Monga contributed equally with all other contributors.

A. S. Thind (✉)
School of Medicine, University of Wollongong, Wollongong, NSW, Australia

llawarra Health & Medical Research Institute, Wollongong, NSW, Australia
e-mail: athind@uow.edu.au

K. Kaur (✉)
National Agri-Food Biotechnology Institute (NABI), Department of Biotechnology, Ministry of Science and Technology (Government of India), Mohali, Punjab, India

I. Monga (✉)
Department of Dermatology, Columbia University Medical Center (CUMC), New York, NY, USA

S. Singh (ed.), *Machine Learning and Systems Biology in Genomics and Health*,
https://doi.org/10.1007/978-981-16-5993-5_3

of lncRNAs. In this chapter, we described the current tools and computational repositories in the area of lncRNA biology covering numerous aspects like lncRNA identification, annotation, role in disease- or cell-type-specific expression, and detailed literature on their recent novel offbeat functions.

3.1 Introduction

It's been almost two decades since Andrew Fire and Craig C. Mello discovered the phenomenon of gene silencing of the complementary target gene (*unc-22*) by exogenous double-stranded RNA (dsRNA) molecules in *Caenorhabditis elegans* and termed it as RNA interference (RNAi) (Fire et al. 1998; Montgomery et al. 1998). Since then, the molecular biology field has witnessed many noncoding RNA (ncRNA) molecules shedding light on an unknown dimension of the complex gene regulatory layers in the genome (Timmons and Fire 1998; Wagner and Sun 1998), eventually leading to the most ambitious dream in molecular biology: sequencing the entire human genome. Before completing the human genome project, it was assumed that only 1.4% of the mammalian genome codes for protein-coding genes and rest were termed "junk DNA" (The ENCODE Project Consortium 2012). Moreover, RNA was merely seen as the messenger of the genetic information in the central dogma of molecular biology. However, after the RNAi discovery and decoding of the human genome project (The ENCODE Project Consortium 2012; Venter 2001), it is well appreciated that more than 90% of the mammalian genome is involved in active transcription, termed *pervasive transcription* (The ENCODE Project Consortium 2007; Greally 2007; Check 2007). Thus, it was envisioned that so-called junk DNA exhibiting active transcription must have other critical regulatory functions (The ENCODE Project Consortium 2007; Check 2007).

The subsequent post-ENCODE era is filled with the novel revelations of various ncRNA molecules of variable length using next-generation sequencing (NGS) and transcriptome sequencing methods (Siggens and Ekwall 2014; Simonti and Capra 2015). These ncRNA molecules exhibit a variety of functions, namely, transcriptional (Li 2015; Djebali 2012), posttranscriptional, epigenetic (Siggens and Ekwall 2014; Eckhardt 2004), and gene regulation via long-range chromatin interactions (Thurman 2012; Rye 2011; Barski 2007; Chen et al. 2014) under a common functional domain termed gene regulation. Thus different regulatory functions of ncRNAs created a panorama of possible functions and created a plausible reason for the pervasive transcription phenomena. Hence, what was considered "junk" came out as treasure playing immense roles in the regulation of various biological processes (Djebali 2012; Kapranov 2007). With these landmark discoveries, the scientific community recognized that the cellular economy is transacted via both protein coding and noncoding genes (Derrien et al. 2012).

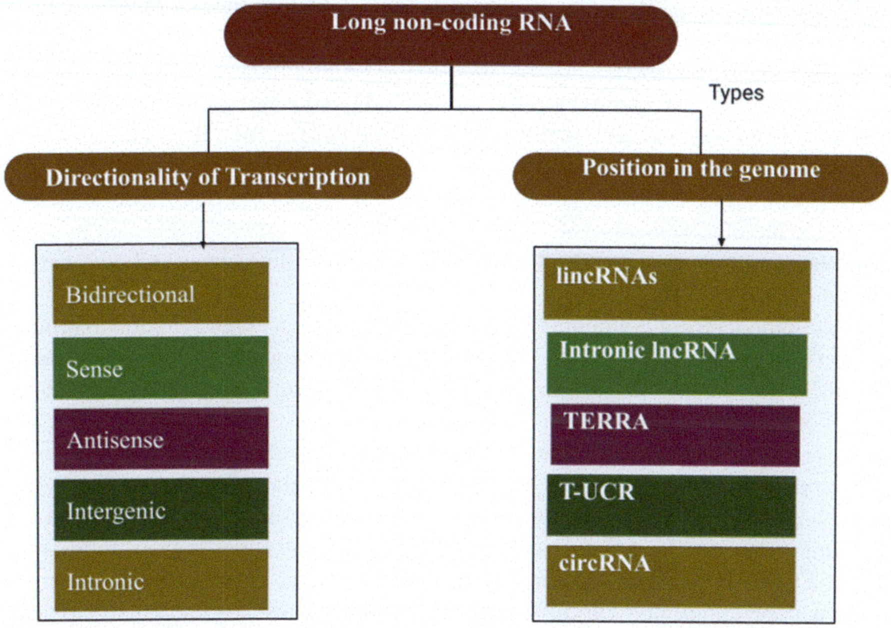

Fig. 3.1 Overview of the categories of long noncoding RNAs (lncRNAs)

Among all ncRNA molecules, lncRNA is the recently discovered, diverse yet most versatile, and pervasively transcribed component of the genome playing as a critical regulatory molecule in almost all the gene regulatory processes of a cell (Derrien et al. 2012). It is the widest category of ncRNAs with a length greater than 200 bp to a few KBs of sequence (Guttman 2009). The lncRNAs exhibit features like transcription by RNA pol-II and pol-III, lack of an open reading frame, canonical polyadenylation signals, splicing, etc. However not all lncRNAs exhibit these explicit signatures (Dahariya et al. 2019), for example, RNA-pol III transcribes some lncRNAs due to their promoter structure (Kung et al. 2013). Several reports suggest that lncRNAs are under a low degree of evolutionary pressure, unlike protein-coding genes.

The LncRNAs are roughly classified based on their position, the direction of transcription (relative to protein-coding genes) and their shape (Laurent et al. 2015; Ma et al. 2013). These are long intergenic noncoding RNAs (lincRNAs): lncRNA transcribed between the genes/exons; long intronic noncoding RNAs (lncRNAs): lncRNA from the intronic regions; telomere-associated ncRNAs (TERRAs): lncRNAs from the telomere region; transcribed ultraconserved regions (T-UCRs); antisense lncRNAs; and circular RNAs (circRNAs) (Fig. 3.1).

The lincRNAs are intergenic and found within the genomic region amid two genes or two exons and possess length from several hundred to thousands of bases. lincRNAs are the most abundant class of lncRNAs. Owing to the advancements in next-generation sequencing technologies, the number of reported lncRNAs has been increasing every year. Recently, around 30,000 and 60,000 human lncRNAs have been reported in different lncRNA annotation projects, and this number is continuously growing (Iyer 2015). They are reported to be involved in various regulatory processes, namely, development (Hon et al. 2017) and cancer. On the other hand, lncRNAs from the intronic regions of protein-coding genes are relatively less abundant as compared to lincRNAs. These are evolutionary conserved long intronic ncRNA sequences, and little is known about their biogenesis and mode of action except for the involvement of RNA pol-II in their transcription as revealed by their coordinated co-expression with corresponding protein-coding genes and presence of poly (A+)-tail.

The telomere repeat-containing RNAs (TERRAs) play a major function to protect the double-strand DNA (dsDNA) ends of linear chromosomes from DNA repair machinery from being recognized as double-stranded cuts. For a long time, they were thought to be transcriptionally inactive because of their protective function. However, it has been reported that some regions in telomere have the potential to be transcribed into ncRNAs termed as telomere repeat-containing RNAs (TERRAs). These are ~100 bp to >9 kb in length and well conserved among eukaryotes including humans. These are usually synthesized by RNA pol-II from the C-rich strand. TERRA transcripts can be found in different cell cycle stages in different amounts depending upon various factors like cellular stress, developmental phase, length of telomere, and telomere chromatin structure (Isken and Maquat, 2009).

T-UCRs are highly conserved sequences of more than 200 bps. In 2004, 481 UCRs were discovered to have complete identity across mouse, rat, and human genomes. These were also >95% identical in other vertebrate genomes. Their altered expression is associated with different diseases like cancer. The circular (CircRNAs) is the most recent identified form of lncRNAs by length, and they are the excellent competing endogenous RNAs (ceRNAs) acting as miRNA sponge to compete with the miRNAome of the host cell and allow for gene expression (Ebert and Sharp 2010).

Additionally, a newer kind of RNAs is discovered with the identification of the pervasive and active transcription at the enhancer regulatory DNA termed as enhancer RNAs (eRNA) (Pennacchio et al. 2013; Quinn and Chang 2016; Andersson et al. 2014). It has been experimentally validated that lncRNA expression was higher at enhancer sites, which exhibit certain chromatin-based features like a high ratio of histone H3 lysine 4 monomethylation to trimethylation (H3K4me1)/(H3K4me3), acetylation of H3K27 (H3K27Ac), and binding motifs for the coactivators (Pennacchio et al. 2013; Li et al. 2016; Medina-Rivera et al. 2018). The eRNA plays their role in gene expression modulation via long-range chromatin interactions and impacts the 3D genome structure in the disease progression. ERNA discovery realized the acceptance of the finding about the active transcription of regulatory DNA as ncRNA, for example, enhancer sequences are not only motif for the binding of enhancer transcription factors, but in fact, it also transcribes the eRNAs, a type of lncRNA (Visel et al. 2009).

The lncRNAs play crucial roles in the continuum of biological processes like protein expression modulation (Costa et al. 2010), alternative splicing (Tripathi et al.

2010), protein localization (Batista and Chang 2013; Hombach and Kretz 2016), and epigenetic regulation (Rinn 2014). Mostly all of the functions performed by lncRNA and its allied types comes under two categories: their regulations in cis or trans. The cis-regulation refers to the regulation of the gene from same strand, while trans conveys its opposite definition.

The cis-regulation involves the three mechanisms (reviewed in detail (Kopp and Mendell 2018; Engreitz et al. 2016). (1) First staple of methods assumes that the act of lncRNA transcription itself is the phenomenon to achieve the gene expression regulation of neighboring genes, because lncRNAs exhibit potential to recruit various regulatory factors to the locus under control and hence modulate of fine-tune their function; one of the historical example of this mechanism is the XIST lncRNA transcription (Cerase et al. 2015). (2) Second model proposed that the process of lncRNA transcription followed by splicing of the lncRNA confers a gene-regulation functionality. This mechanism is very well detailed illustrated and validated in historical discovery of lncRNA Airn (Latos et al. 2012). (3) Third mechanism assumes that major function of the regulation in cis chiefly depends on DNA elements also known as motifs near or at the lncRNA promoter or gene locus. The role of DNA elements or motifs in lncRNA locus and gene locus was illustrated in the case of *lincRNA-p21.* lincRNA-p21 is a nuclear-localized transcript transcribed from an intergenic locus in the vicinity of the CDKN1A gene (Huarte et al. 2010). However, on the other hand, trans-regulatory mechanisms involve the gene expression regulation on the other strand and sometimes far away from the lncRNA locus. The trans gene regulatory mechanisms usually operate via chromatin looping.

In the present chapter, we systematically provided the updated collection of the computational resources, databases, and tools in the lncRNA biology. We also covered the up-to-date novel concepts in the lncRNA field. The focus of the present chapter is to provide a bird's-eye view of the tools, databases pertaining to lncRNAs as of 2021.

3.2 lncRNAs Analysis

3.2.1 Databases Harboring lncRNAs

Major projects covering lncRNA discovery and main catalogues

1. FANTOM
2. Gencode
3. GTEx
4. NONCODE

3.2.1.1 General Databases for lncRNA Storage and Annotation

Many efforts have been put to organize the information on the data of lncRNAs. This information is basically catalogued in databases with basis of genomic annotation, expression profiling, molecular interactions, and sequence variants. A detailed overview of databases harboring information regarding lncRNAs is discussed in the section below.

1. GENCODE: GENCODE is a large consortium dedicated to annotate validated gene features in the human genome. It cartels computational analysis, manual curation, and validation of GENCODE transcript database. At present it harbors 17,910 lncRNA genes along with 48,351 transcripts in humans and 13,201 genes and 18,800 transcripts in mice (Harrow et al. 2012).
2. LNCipedia: LNCipedia is an annotated database of lncRNA transcripts from humans. It encompasses 127,802 lncRNA transcripts and 56,946 genes, which are annotated and obtained from different sources. In addition to this database, it also contains information regarding gene structure like secondary structure and binding sites of miRNAs. This also highlights that the lncRNAs have secondary structures just like miRNAs and have strong association with proteins. The depository is also equipped with a tool for calculating coding potential of protein with coding potential calculator and HMMER. It is freely available for public use and can be used to search and download lncRNA transcripts (http://www.lncipedia.org) (Volders et al. 2013).
3. lncRNAdb: This database harbors details of lncRNA in eukaryotes which are annotated. Individual entry contains information regarding sequence of RNA, genomic context, expression, structure, and subcellular localization. Furthermore, lncRNAdb is linked with UCSC genome browser for easy visualization. This database has been updated and contains 287 lncRNAs from eukaryotes and can be explored by using the link http://lncrnadb.org (Quek et al. 2015; Amaral et al. 2011).
4. LncRNASNP2: This is a comprehensive database designed to hold information on single nucleotide polymorphism (SNP) in mouse and human lncRNAs. It contains details of lncRNA-miRNA binding, effect of SNPs on the structure of lncRNAs, and mutations in lncRNAs. lncRNASNP2 have 141,353 and 10,205,295 lncRNAs and SNPs in total and also incorporate cosmic noncoding variations (859,534) as well as their effect on the structure of lncRNAs. Additionally, it has data related to TCGA cancer mutations (315,234) and their impact on lncRNAs. This database is entrenched with various analysis tools for new variants in lncRNA and is freely accessible at http://bioinfo.life.hust.edu.cn/lncRNASNP2 (Miao et al. 2018).
5. LncBook: LncBook is a curated repository of long noncoding RNAs in humans. This database has a systematic curation of lncRNAs by multi-omics data amalgamation, association of disease, and functional annotation. It houses around 270,044 lncRNAs along with 1867 functional lncRNAs and 3762 lncRNA functional associations. Apart from this, LncBook integrates multi-omics data from methylation, genomic variation, and miRNA-lncRNA interactions. Also, there is information regarding experimentally validated disease associated lncRNAs in humans. LncBook is easily accessible at http://bigd.big.ac.cn/lncbook (Ma et al. 2019).

6. LncBase: This database gives a comprehensive annotation of lncRNA:miRNA interactions. It has both experimental (>5000) and predicted lncRNA (>ten million) miRNA recognition elements on mouse and human lncRNA. This resource is implemented with options like graphical visualization of genomic locations of transcripts, their binding sites, etc. LncBase is updated, and more entries have been incorporated to make it more comprehensive with >70,000 miRNA:LncRNA interactions which are experimentally validated. This database can be accessed by https://diana.e-ce.uth.gr/lncbasev3 (Paraskevopoulou et al. 2013).
7. NONCODE: NONCODE is an instructive and valuable resource of noncoding RNAs from animals and plant species. Currently it harbors 39 species (23 plants and 16 animals) and 527,336 lncRNA transcripts and 337, 880 lncRNA genes from 16 species in totality. This can be explored by using http://www.noncode.org/ (Liu et al. 2005).

3.2.1.2 Databases for Long Noncoding RNA Expression

1. The Noncoding RNA Expression Database (NRED): NRED is a public repository, which provides information on gene expression of lncRNAs in humans and mice. This database contains data from microarray and in situ hybridization. It includes three datasets, namely, custom noncoding microarray, GNF SymAtlas, and Allen Brain Atlas. Among them GNF SymAtlas have 5692 mouse and 1287 human probes exclusively documented as long ncRNAs. Out of them 3403 were expressed in mouse and 733 in humans. This depository is available at http://jsm-research.imb.uq.edu.au/NRED and is implemented with easy browsing facilities (Dinger et al. 2009).
2. lncRNAKB: lncRNAKB is a cohesive database for exploring lncRNAs on the basis of tissue specificity and disease association. It contains 77,199 lncRNAs from humans collected from six independent datasets. This repository is a compilation of expression patterns of lncRNAs derived from RNA-seq data and identification of co-expression modules through network analysis and pathway enrichment. Further, data related to expression quantitative trait loci obtained from whole genome sequence genotype data can also be downloaded through this resource. This includes analyzing phylogenetic conservation, tissue specificity, and coding potential calculation. LncRNAKB is accessible via http://www.lncrnakb.org/, and associated data are freely obtainable through Open Science Framework (https://doi.org/10.17605/OSF.IO/RU4D2) (Seifuddin et al. 2020).
3. LncExpDB: LncExpDB is an expression database of human lncRNAs devoted to give expression profiles of lncRNA genes along with their functions, interactions with protein-coding genes through diverse biological conditions. This repository currently has expression profiles of 101,293 human lncRNA genes resulting from 1977 samples across 337 biological conditions. It is also having details of 28,443,865 mRNA:lncRNA interactions along with a user-friendly web server

for easy exploration and visualization of data. LncExpDB can be explored on https://bigd.big.ac.cn/lncexpdb/ (Li et al. 2021).

3.2.1.3 Specialized lncRNA Databases

Disease-Specific Databases

1. LncTarD: This is a manually created repository of information on lncRNA-influenced functions and lncRNA-mediated regulatory mechanisms involved in human diseases. LncTarD contains 2822 key lncRNA–target regulations encompassing 475 lncRNAs and 1039 targets coupled with 177 human diseases. Moreover, it also has 1613 experimentally supported functional regulations and 1209 expression relations among human diseases. LncTarD also offers an interactive interface for convenient browsing, searching, and downloading facility. The webserver can be accessed via http://biocc.hrbmu.edu.cn/LncTarD/ (Zhao et al. 2020).
2. LncRNADisease: This database is dedicated to have clear understanding about the diseases associated with lncRNAs due to their dysfunction. The LncRNADisease database houses 487 entries of lncRNA, which are associated with diseases. The compendium has details of 166 diseases and other molecules that interact with lncRNAs at molecular levels like DNA, RNA, proteins, and miRNAs. Moreover, it is further linked with lncRNA-disease associations with genomic information, species, sequences, and references. A detailed description for each entry is provided, and the database is further linked with a bioinformatic method to predict novel lncRNA-disease associations and added that entries also. LncRNADisease was updated in 2019 as LncRNADisease v2.0 and contains 19,166 entries of lncRNA, and ~ 200,000 lncRNA-disease associations are integrated in the main database. The database is freely available at http://www.rnanut.net/lncrnadisease/ (Chen et al. 2013; Bao et al. 2019).
3. CRlncRNA: CRlncRNA is a manually constructed and curated repository of lncRNA in cancer. All the entries provided here and experimentally validated. The existing version of CRlncRNA harbors 355 entries of cancer-specific lncRNAs, covering 1072 cancer-lncRNA associations considering 76 types of cancer and further 1238 interactions with diverse RNAs and proteins. This repository can be explored by the link http://crlnc.xtbg.ac.cn./ (Wang et al. 2018).
4. CVDncR: CVDncR is a database harboring data of noncoding RNAs like lncRNAs and miRNA involved in cardiovascular diseases. The database includes entries of around 23 cardiovascular diseases with their detailed information. Currently, the database has 99 lncRNAs, 426 miRNAs, and 24 circular RNAs and is accessible via http://www.sysbio.org.cn/cvdncr/About%20Us.html (Wu et al. 2020).
5. TANRIC: Database of noncoding RNAs in cancer (Li et al. 2015).

Organism-Specific Databases

1. PLncDB: This database is a compilation of 12,46,372 lncRNAs from 80 plant species. The data is mainly extracted from 13,834 RNA-seq libraries and contains

details of lncRNA targets and their regulatory networks. It is freely available at http://plncdb.tobaccodb.org/ (Jin et al. 2021; Jin et al. 2013).

2. CANTATAdb 2.0: CANTATAdb2.0 is a repository of lncRNA in plants and algae, which are identified computationally from 36 plant species and three algae. Currently, it encompasses 239,631 lncRNAs mined from RNA-seq data, and information like expression and coding potential has been integrated in this database. Its older version, that is, CANTATAdb, had only ten species. The data are freely accessible for browsing, searching, and downloading at http://yeti.amu.edu.pl/CANTATA/ (Szcześniak et al. 2019).
3. EVLncRNAs: It is a manually curated database of lncRNAs in plants validated by experimental studies. The current version of EVLncRNAs includes 1543 lncRNAs from 77 species that is three times higher than the existing experimentally validated databases. The database is accessible via http://biophy.dzu.edu.cn/EVLncRNAs (Zhou et al. 2018; Zhou et al. 2021) .
4. GREENC: This database contains lncRNA specific to plants and algae. GREENC has ~120,000 annotated lncRNAs from 37 plant species and six algae. The database is available at http://greenc.sciencedesigners.com/ (Paytuví Gallart et al. 2016).

Databases on Subcellular Localization of Long Noncoding RNAs

1. LncSLdb: This is a repository that has information regarding the subcellular localization information of lncRNAs. The current version of this database has >11,000 transcripts and 9494 lncRNA genes from human, mouse, and fruit fly. Based on the accumulation pattern of lncRNAs, they have been categorized into nucleus, cytoplasm, or nucleus/cytoplasm. Moreover apart from downloading and browsing, this resource also provides a comprehensive list of tools in order to search genomic coordinates, gene symbols, and sequence similarity. LncSLdb can be browsed on http://bioinformatics.xidian.edu.cn/lncSLdb (Wen et al. 2018).
2. LncATLAS: LncATLAS is a comprehensive repository of localization of human lncRNA in various compartments of a cell from RNA-seq data. It encompasses 6768 lncRNAs annotated from GENCODE and distributed among different compartments across 15 cell lines. LncATLAS is available through an instinctive and instructive web server, in which lncRNAs can be accessed using names or identifiers. This is a relational database that is accessible through lncatlas.crg.eu (Mas-Ponte et al. 2017).

3.2.2 Tools and Algorithms for lncRNA

Bioinformatics tools help the researchers discover new lncRNA and characterize the lncRNA for its functionality and localization. Obtaining the in silico results beforehand could help in guiding experimental setup. Identifying novel lncRNA transcripts and predicting novel transcripts' coding potential are important because it helps annotate lncRNA accurately. In earlier times, the in vitro methods were frequently employed to verify the novel transcripts' coding potential, but those methods are time-consuming, especially for large numbers of transcripts (Pinkney et al. 2020). Several computational predictive tools are developed to predict the transcripts' protein-coding potential. However, each tool has its criteria for the prediction of the protein-coding potential, but in general most of the algorithms depend on homology sequence similarity. The protein-coding potential and the understanding of lncRNAs' potential function helps to unreveal the cellular level complexity of the biological system. The subcellular localization predictions help, to some extent, understand lncRNA potential function. Bioinformatics tools for the lncRNA can be categorized into five types based on their functional role: identification of lncRNA, protein-coding potential prediction, subcellular localization, structural conformation, and interaction and function prediction. These tools are summarized in Table 3.1.

3.3 Circular RNA Analyses

Like the linear transcripts, many studies have recently reported the back spliced (circular) transcripts' functional impacts and their coding potential. With the improved bioinformatics methods, it is possible to identify novel circular RNA from NGS data. However, this field is recent, but significant progress has been made in the bioinformatics method development and the circular RNA's characterization. Differential expression analyses are a widely used approach to investigate the transcriptomic changes in various disease conditions or cellular states. Just like gene differential expression analyses, identification, quantification, characterization, and functional interpretation are the important steps for differential expression analyses of circular RNA from RNAseq. List of various tools used for these steps is mentioned in Table 3.1.

All identification tools depend upon external aligners with BWA, Bowtie, and STAR as the common preferences. They can be further classified based on the algorithm basis, that is, BSJ based, machine learning based, and integrated approaches (Chen et al. 2021). Quantification is the scale used to measure the levels of expression of circular RNA and is one of the basic steps required for the differential expression analyses to measure the difference in the expression among different phenotypic or disease states of the organisms. Many tools which can perform both tasks, that is, identification and quantification.

1. Backsplice Junction Based (BSJ): Identification of backsplice junction is the key feature for discovering circRNA, and most of the tools are based on the BSJ approach. These algorithms either use predefined BSJ and flanking sequences or work based on segmented-read maps.
2. Machine Learning-Based circRNA Identification Tools: These tools use the features and knowledge of real circRNA to train and learn the ML model for the classification. Alu repeats and structural and sequence motifs are the most

Table 3.1 Description of the tools related to various categories for LncRNAs

Tool name	Description	Year	Platform/ programming environment	Link/ reference
RNAsamba	lncRNA identification based on neural network (all species)	2020	Python/Docker	Camargo et al. (2020)
Lncident	Identification of lncRNA utilizing sequence intrinsic composition and open reading frame information	2016	R package and web server	Han et al. (2016)
CPC2	Protein coding potential of non-coding RNA based on sequence intrinsic features	2017	Web based	Kang et al. (2017)
CPPred	Coding potential prediction based on SVM to distinguish ncRNAs from coding RNAs using sequence features, such as ORF length, ORF coverage, ORF integrity, Fickett score, Hexamer score, PI, Gravy, Instability index and CTD features	2019	Python based	Tong and Liu (2019)
CNIT	Identification of protein-coding potential for Long non-coding transcripts based on intrinsic sequence composition	2019	Web tool	Guo et al. (2019)
SubMito-XGBoost	Submitochondrial proteins subcellular localization	2020	R-based	Yu et al. (2020)
LncLocator	Subcellular localization predictor for long non-coding RNAs based on a stacked ensemble classifier	2018	Web based	Cao et al. (2018)
SPOT-RNA	RNA secondary structure prediction using an ensemble of two-dimensional deep neural networks and transfer learning	2019	Webserer/ Python-based tested on unix/ linux like environment	Singh et al. (2019)
SCLpred-EMS	Subcellular localization prediction of the endomembrane system and secretory pathway proteins by deep N-to-1 convolutional neural networks	2020	Web based	Kaleel et al. (2020)
DeepLoc	Prediction of protein subcellular localization using deep learning	2017	Web based	Almagro Armenteros et al. (2017)

(continued)

Table 3.1 (continued)

Tool name	Description	Year	Platform/ programming environment	Link/ reference
Tools specialized for CircularRNA				
Mapsplice	Identification/quantification of circRNA (tested using real or simulated datasets)	2016 (v2.2.1)	Linux/Unix	Wang et al. (2010)
Segemehl	Identification of circRNA	2018	Linux/Unix	Hoffmann et al. (2014)
KNIFE	Identification of circRNA based on BSJ and use logistic generalized linear model (tested using real or simulated datasets)	2015	Linux/Unix	Szabo et al. (2015)
circRNA finder	Identification/quantification of the circRNA from BSJ (tested using real or simulated datasets)	2016	Linux/Unix	Fu and Liu (2014)
CIRCexplorer	Identification/quantification of circRNA based on BSJ (tested using real or simulated datasets)	2019 (V2.3.8)	Linux/Unix	Zhang et al. (2016)
PTESFinder	Identification/quantification of circRNA based on BSJ	2016	Linux/Unix	Izuogu et al. (2016)
UROBORUS	Identification/quantification of circRNA based on BSJ	2018	Linux/Unix	Song et al. (2016)
find_circ	Identification/quantification of circRNA based on BSJ	2015 (v1.2)	Linux/Unix	Memczak et al. (2013)
Circtools	Identification/quantification of circRNA based on BSJ (harmonized pipeline covers different steps of in silico circRNA analyses, from prediction to functional analyses)	2019 (1.1.0.8)	Linux/Unix	Jakobi et al. (2019)
CIRI2	Identification/quantification BSJ based	2017	Linux/Unix	Gao et al. (2018)
DCC and CircTest	Identification/quantification BSJ based	2019	Linux/Unix	Cheng et al. (2016)
PredcircRNA	Identification ML based	2015	Linux/Unix	Pan and Xiong (2015)
WebCircRNA	Identification ML based stem cell-specific circRNA	2018	Web, Linux/Unix	Pan et al. (2018)
DeepCirCode	Identification CNN (deep learning) based human circRNA	2019	Linux/Unix/ window	Wang and Wang (2019)

(continued)

Table 3.1 (continued)

Tool name	Description	Year	Platform/ programming environment	Link/ reference
CIRIquant	Accurate quantification and differential expression analysis of circRNAs	2020	Linux/Unix	Zhang et al. (2020)
CIRI-full	Combine reverse overlap (RO) and BSJ features for reconstruction of full-length circRNAs and isoform-level quantification	2019	Linux/Unix	Zheng et al. (2019)
Sailfish-cir	Quantification of circRNAs using model-based framework	2017	Unix-like environment	Li et al. (2017)
Co-expression module detection				
WGCNA	Co-expression network module detection using hierarchical clustering	2008	Linux/Unix/ window	Langfelder and Horvath (2008)
DiffCoEx	Identify and group differentially co-expressed genes	2010	R package	Tesson et al. (2010)
DICER	Identifies modules that correlate differently between sample groups	2013	Java	Amar et al. (2013)
DINGO	Groups genes based on differential behavior in a particular condition/ phenotypic state	2015	R package	Ha et al. (2015)
GSCNA	Multivariate differential co-expression test for gene sets	2014	R package	Rahmatallah et al. (2014)
Biclustering	Identify unique modules to a subpopulation of samples without the prior knowledge of the group of samples	2015	Method	Pontes et al. (2015)
Regulatory network inference				
Genie3	Include transcription factors (TF) information to construct a regulatory network by determining the TF expression pattern	2010	R/Bioconductor	Huynh-Thu et al. (2010)\
ARACNE	Only edges which are expected to be regularity are retained after filtering of indirect connections between genes	2014	Cytoscape (Unix/ Linux/window)	Chávez Montes et al. (2014)
Cytoscape	The visualization of networks	2003	Linux/Unix/ window	Shannon et al. (2003)

common feature used for the ML model in circRNA identification, distinguishing circRNA from lncRNA and predicting circRNA.
3. Other Integrated Tools: These tools help reduce the false-positive prediction of the circular RNA identification. These approaches usually merge the results from two and more tools.

3.4 Co-Expression Regulatory Network Prediction

Studying gene regulation through co-expression or gene regulatory networks could play an important role in building novel hypotheses by predicting the potential interactors in the biological or metabolic pathway/s. Genes with similar expression patterns tend to have similar functions, and these co-expressed genes are usually involved in the same metabolic pathway. So the study of the co-expressed gene has the potential to reveal the unknown gene regulatory interactions. This network analysis can be further categorized into the gene co-expression network (GCN); gene regulatory network (GRN) form co-expression network analyses. GCN network doesn't provide any information about the direction; on the other hand, the GRN network is a directed network representing various biochemical processes. Analyses can help look for activators or inhibitors of a specific target in the pathways or for the pathway in itself. Co-expression can be determined between gene-gene and gene-lncRNA, and there are various types of co-expression methods proposed, including Pearson's correlation coefficient, mutual information, Spearman's rank correlation coefficient, Kendall correlation, and Euclidean distance. Each co-expression coefficient has its advantages and disadvantages. Pearson's correlation is a popular choice, but outliers can easily influence the outcomes when samples are few. On the other hand, Spearman's coefficient can handle outliers more efficiently for the same conditions because it tackles the outliers using the rank approach on the sorted gene expression values (Schober et al. 2018).

To calculate the correlation, R and Cytoscape platforms are widely used for network analyses. "Corr" and "cor.test" in R are used for co-expression and P-value calculations, respectively. The "psych" R package comes with inbuilt features of p-values and FDR calculation, and "corplot" library is a network visualization R package. ExpressionCorrelation is a Cytoscape application which makes similarity networks. Further, the addition of clustering approaches to the network helps group the genes with similar expression patterns in multiple samples, and cluster modules could be phenotype-specific or may represent some biological process. Weighted gene correlation network analysis (WGCNA) (Zhang and Horvath 2005) is a widely used clustering package which uses the hierarchical approach. Other packages such as ARACNE (Margolin et al. 2006) and GENIE3 (Huynh-Thu et al. 2010) are designed to infer causality by constructing regulatory networks. Co-LncR NA (Zhao et al. 2015) tool is specifically designed to identify co-expressed protein-coding genes of lncRNAs that can shed light into the lncRNA functions.

3.5 Conclusion and Future Aspects

With the advent of the NGS and other multi-omics-based advancements, a large number of lncRNAs have been identified. Despite these advancements, understanding of the multifarious functions of these immensely important regulatory molecules, their diverse and complex biological and molecular functions remain to be determined. Keeping view of their immense regulatory roles, their aberrant expression is reported in many cellular perturbations or disease conditions like viral diseases, infections, autoimmune, and complex genetic disorders like cancers. Thus there is an urgent need to develop comprehensive computational tools, databases, and algorithms to better understand the regulatory potential of lncRNAs. The current chapter has provided an overview of the present resources in the field of lncRNAs. The more understanding in lncRNA mechanism will be helpful to enable their use in the development of lncRNA-based therapeutics as it already witnessed promising lncRNA-based antivirals against viruses including SARS-coV2 and chikungunya virus which have been reported (Hucke and Bugert 2020). In nutshell, the advancements in experimental approaches coupled with the computational resources will assist the noncoding genomics fraternity to unlock and utilize the potential of ncRNAs against various diseases.

References

Almagro Armenteros JJ et al (2017) DeepLoc: prediction of protein subcellular localization using deep learning. Bioinformatics 33(21):3387–3395

Amar D, Safer H, Shamir R (2013) Dissection of regulatory networks that are altered in disease via differential co-expression. PLoS Comput Biol 9(3):e1002955

Amaral PP et al (2011) lncRNAdb: a reference database for long noncoding RNAs. Nucleic Acids Res 39:D146–D151

Andersson R et al (2014) An atlas of active enhancers across human cell types and tissues. Nature 507(7493):455–461

Bao Z et al (2019) LncRNADisease 2.0: an updated database of long non-coding RNA-associated diseases. Nucleic Acids Res 47(D1):D1034–d1037

Barski A et al (2007) High-resolution profiling of histone methylations in the human genome. Cell 129(4):823–837

Batista PJ, Chang HY (2013) Long noncoding RNAs: cellular address codes in development and disease. Cell 152(6):1298–1307

Camargo AP et al (2020) RNAsamba: neural network-based assessment of the protein-coding potential of RNA sequences. NAR Genom Bioinform 2(1):lqz024

Cao Z et al (2018) The lncLocator: a subcellular localization predictor for long non-coding RNAs based on a stacked ensemble classifier. Bioinformatics 34(13):2185–2194

Cerase A et al (2015) Xist localization and function: new insights from multiple levels. Genome Biol 16(1):166

Chávez Montes RA et al (2014) ARACNe-based inference, using curated microarray data, of Arabidopsis thaliana root transcriptional regulatory networks. BMC Plant Biol 14:97

Check E (2007) Genome project turns up evolutionary surprises. Nature 447:760–761

Chen G et al (2013) LncRNADisease: a database for long-non-coding RNA-associated diseases. Nucleic Acids Res 41:D983–D986

Chen L et al (2021) The bioinformatics toolbox for circRNA discovery and analysis. Brief Bioinform 22(2):1706–1728

Chen P, Wang Y, Li GJN (2014) Dynamics of histone variant H3. 3 and its coregulation with H2A. Z at enhancers and promoters. Nucleus 5(1):21–27

Cheng J, Metge F, Dieterich C (2016) Specific identification and quantification of circular RNAs from sequencing data. Bioinformatics 32(7):1094–1096

Costa PR, Acencio ML, Lemke N (2010) A machine learning approach for genome-wide prediction of morbid and druggable human genes based on systems-level data. BMC Genomics 11:S9

Dahariya S et al (2019) Long non-coding RNA: Classification, biogenesis and functions in blood cells. Mol Immunol 112:82–92

Derrien T et al (2012) The GENCODE v7 catalog of human long noncoding RNAs: analysis of their gene structure, evolution, and expression. Genome Res 22(9):1775–1789

Dinger ME et al (2009) NRED: a database of long noncoding RNA expression. Nucleic Acids Res 37:D122–D126

Djebali S et al (2012) Landscape of transcription in human cells. Nature 489(7414):101–108

Ebert MS, Sharp PA (2010) MicroRNA sponges: progress and possibilities. RNA 16(11): 2043–2050

Eckhardt F (2004) et al. Future potential of the human epigenome project 4(5):609–618

Engreitz JM et al (2016) Local regulation of gene expression by lncRNA promoters, transcription and splicing. Nature 539(7629):452–455

Fire A et al (1998) Potent and specific genetic interference by double-stranded RNA in Caenorhabditis elegans. Nature 391(6669):806–811

Fu X, Liu R (2014) CircRNAFinder: a tool for identifying circular RNAs using RNA-Seq data. in Proceedings of the 6th International Conference on Bioinformatics and Computational Biology, BICOB

Gao Y, Zhang J, Zhao F (2018) Circular RNA identification based on multiple seed matching. Brief Bioinform 19(5):803–810

Greally JM (2007) Encyclopaedia of humble DNA. Nature 447(7146):782–783

Guo JC et al (2019) CNIT: a fast and accurate web tool for identifying protein-coding and long non-coding transcripts based on intrinsic sequence composition. Nucleic Acids Res 47(W1):W516–W522

Guttman M et al (2009) Chromatin signature reveals over a thousand highly conserved large non-coding RNAs in mammals. Nature 458(7235):223–227

Ha MJ, Baladandayuthapani V, Do KA (2015) DINGO: differential network analysis in genomics. Bioinformatics 31(21):3413–3420

Han S et al (2016) Lncident: a tool for rapid identification of long noncoding RNAs utilizing sequence intrinsic composition and open reading frame information. Int J Genomics 2016:9185496

Harrow J et al (2012) GENCODE: the reference human genome annotation for the ENCODE project. Genome Res 22(9):1760–1774

Hoffmann S et al (2014) A multi-split mapping algorithm for circular RNA, splicing, trans-splicing and fusion detection. Genome Biol 15(2):R34

Hombach S, Kretz M (2016) Non-coding RNAs: classification, biology and functioning. Adv Exp Med Biol 937:3–17

Hon CC et al (2017) An atlas of human long non-coding RNAs with accurate 5′ ends. Nature 543(7644):199–204

Huarte M et al (2010) A large intergenic noncoding RNA induced by p53 mediates global gene repression in the p53 response. Cell 142(3):409–419

Hucke FIL, Bugert JJ (2020) Current and promising antivirals against chikungunya virus. Front Public Health 8:618624

Huynh-Thu VA et al (2010) Inferring regulatory networks from expression data using tree-based methods. PLoS One 5:9

Iyer MK et al (2015) The landscape of long noncoding RNAs in the human transcriptome. Nat Genet 47(3):199–208
Izuogu OG et al (2016) PTESFinder: a computational method to identify post-transcriptional exon shuffling (PTES) events. BMC Bioinformatics 17:31
Jakobi T, Uvarovskii A, Dieterich C (2019) Circtools-a one-stop software solution for circular RNA research. Bioinformatics 35(13):2326–2328
Jin J et al (2013) PLncDB: plant long non-coding RNA database. Bioinformatics 29(8):1068–1071
Jin J et al (2021) PLncDB V2.0: a comprehensive encyclopedia of plant long noncoding RNAs. Nucleic Acids Res 49(D1):D1489–d1495
Kaleel M et al (2020) SCLpred-EMS: subcellular localization prediction of endomembrane system and secretory pathway proteins by deep N-to-1 convolutional neural networks. Bioinformatics 36(11):3343–3349
Kang YJ et al (2017) CPC2: a fast and accurate coding potential calculator based on sequence intrinsic features. Nucleic Acids Res 45(W1):W12–W16
Kapranov P et al (2007) RNA maps reveal new RNA classes and a possible function for pervasive transcription. Science 316(5830):1484–1488
Kopp F, Mendell JT (2018) Functional classification and experimental dissection of long noncoding RNAs. Cell 172(3):393–407
Kung JT, Colognori D, Lee JT (2013) Long noncoding RNAs: past, present, and future. Genetics 193(3):651–669
Langfelder P, Horvath S (2008) WGCNA: an R package for weighted correlation network analysis. BMC Bioinformatics 9:559
Latos PA et al (2012) Airn transcriptional overlap, but not its lncRNA products, induces imprinted Igf2r silencing. Science 338(6113):1469–1472
Laurent GS, Wahlestedt C, Kapranov P (2015) The landscape of long noncoding RNA classification. Trends Genet 31(5):239–251
Li J et al (2015) TANRIC: an interactive open platform to explore the function of lncRNAs in cancer. Cancer Res 75(18):3728–3737
Li M et al (2017) Quantifying circular RNA expression from RNA-seq data using model-based framework. Bioinformatics 33(14):2131–2139
Li MJ et al (2015) Exploring the function of genetic variants in the non-coding genomic regions: approaches for identifying human regulatory variants affecting gene expression. Brief Bioinform 16(3):393–412
Li W, Notani D, Rosenfeld MG (2016) Enhancers as non-coding RNA transcription units: recent insights and future perspectives. Nat Rev Genet 17(4):207–223
Li Z et al (2021) LncExpDB: an expression database of human long non-coding RNAs. Nucleic Acids Res 49(D1):D962–d968
Liu C et al (2005) NONCODE: an integrated knowledge database of non-coding RNAs. Nucleic Acids Res 33:D112–D115
Ma L, Bajic VB, Zhang Z (2013) On the classification of long non-coding RNAs. RNA Biol 10(6): 924–933
Ma L et al (2019) LncBook: a curated knowledgebase of human long non-coding RNAs. Nucleic Acids Res 47(D1):D128–d134
Margolin AA et al (2006) ARACNE: an algorithm for the reconstruction of gene regulatory networks in a mammalian cellular context. BMC Bioinformatics 7:S7
Mas-Ponte D et al (2017) LncATLAS database for subcellular localization of long noncoding RNAs. RNA 23(7):1080–1087
Medina-Rivera A et al (2018) Widespread enhancer Activity from Core promoters. Trends Biochem Sci 43(6):452–468
Memczak S et al (2013) Circular RNAs are a large class of animal RNAs with regulatory potency. Nature 495(7441):333–338
Miao YR et al (2018) lncRNASNP2: an updated database of functional SNPs and mutations in human and mouse lncRNAs. Nucleic Acids Res 46(D1):D276–d280

Montgomery MK, Xu S, Fire A (1998) RNA as a target of double-stranded RNA-mediated genetic interference in Caenorhabditis elegans. PNAS 95(26):15502–15507

Pan X, Xiong K (2015) PredcircRNA: computational classification of circular RNA from other long non-coding RNA using hybrid features. Mol Biosyst 11(8):2219–2226

Pan X et al (2018) WebCircRNA: classifying the circular RNA potential of coding and noncoding RNA. Genes (Basel) 9:11

Paraskevopoulou MD et al (2013) DIANA-LncBase: experimentally verified and computationally predicted microRNA targets on long non-coding RNAs. Nucleic Acids Res 41:D239–D245

Paytuví Gallart A et al (2016) GREENC: a wiki-based database of plant lncRNAs. Nucleic Acids Res 44(D1):D1161–D1166

Pennacchio LA et al (2013) Enhancers: five essential questions. Nat Rev Genet 14(4):288–295

Pinkney HR, Wright BM, Diermeier SD (2020) The lncRNA toolkit: databases and in silico Tools for lncRNA analysis. Noncoding RNA 6:4

Pontes B, Giráldez R, Aguilar-Ruiz JS (2015) Biclustering on expression data: a review. J Biomed Inform 57:163–180

Quek XC et al (2015) lncRNAdb v2.0: expanding the reference database for functional long noncoding RNAs. Nucleic Acids Res 43:D168–D173

Quinn JJ, Chang HY (2016) Unique features of long non-coding RNA biogenesis and function. Nat Rev Genet 17(1):47–62

Rahmatallah Y, Emmert-Streib F, Glazko G (2014) Gene sets net correlations analysis (GSNCA): a multivariate differential coexpression test for gene sets. Bioinformatics 30(3):360–368

Rinn JL (2014) lncRNAs: linking RNA to chromatin. Cold Spring Harb Perspect Biol 6:8

Rye M et al (2011) Clustered ChIP-Seq-defined transcription factor binding sites and histone modifications map distinct classes of regulatory elements. BMC Biol 9(1):1–18

Schober P, Boer C, Schwarte LA (2018) Correlation coefficients: appropriate use and interpretation. Anesth Analg 126(5):1763–1768

Seifuddin F et al (2020) lncRNAKB, a knowledgebase of tissue-specific functional annotation and trait association of long noncoding RNA. Sci Data 7(1):326

Shannon P et al (2003) Cytoscape: a software environment for integrated models of biomolecular interaction networks. Genome Res 13(11):2498–2504

Siggens L, Ekwall K (2014) Epigenetics, chromatin and genome organization: recent advances from the ENCODE project. J Intern Med 276(3):201–214

Simonti CN, Capra JA (2015) The evolution of the human genome. Curr Opin Genet Dev 35:9–15

Singh J et al (2019) RNA secondary structure prediction using an ensemble of two-dimensional deep neural networks and transfer learning. Nat Commun 10(1):5407

Song X et al (2016) Circular RNA profile in gliomas revealed by identification tool UROBORUS. Nucleic Acids Res 44(9):e87

Szabo L et al (2015) Statistically based splicing detection reveals neural enrichment and tissue-specific induction of circular RNA during human fetal development. Genome Biol 16(1):126

Szcześniak MW et al (2019) CANTATAdb 2.0: expanding the collection of plant long noncoding RNAs. Methods Mol Biol 1933:415–429

Tesson BM, Breitling R, Jansen RC (2010) DiffCoEx: a simple and sensitive method to find differentially coexpressed gene modules. BMC Bioinformatics 11:497

The ENCODE Project Consortium (2007) Identification and analysis of functional elements in 1% of the human genome by the ENCODE pilot project. Nature 447(7146):799

The ENCODE Project Consortium (2012) An integrated encyclopedia of DNA elements in the human genome. Nature 489(7414):57

Thurman RE et al (2012) The accessible chromatin landscape of the human genome. Expert Rev Molec Diagn 489(7414):75–82

Timmons L, Fire A (1998) Specific interference by ingested dsRNA. Nature 395(6705):854–854

Tong X, Liu S (2019) CPPred: coding potential prediction based on the global description of RNA sequence. Nucleic Acids Res 47(8):e43

Tripathi V et al (2010) The nuclear-retained noncoding RNA MALAT1 regulates alternative splicing by modulating SR splicing factor phosphorylation. Mol Cell 39(6):925–938
Venter JC et al (2001) The sequence of the human genome. Nature 291(5507):1304–1351
Visel A, Rubin EM, Pennacchio LA (2009) Genomic views of distant-acting enhancers. Nature 461(7261):199–205
Volders PJ et al (2013) LNCipedia: a database for annotated human lncRNA transcript sequences and structures. Nucleic Acids Res 41:D246–D251
Wagner RW, Sun LJN (1998) Functional genomics double-stranded RNA poses puzzle. Nature 391(6669):744–745
Wang J, Wang L (2019) Deep learning of the back-splicing code for circular RNA formation. Bioinformatics 35(24):5235–5242
Wang J et al (2018) CRlncRNA: a manually curated database of cancer-related long non-coding RNAs with experimental proof of functions on clinicopathological and molecular features. BMC Med Genet 11(Suppl 6):114
Wang K et al (2010) MapSplice: accurate mapping of RNA-seq reads for splice junction discovery. Nucleic Acids Res 38(18):e178
Wen X et al (2018) lncSLdb: a resource for long non-coding RNA subcellular localization. Database (Oxford) 2018:1–6
Wu R et al (2020) Phenotype-genotype network construction and characterization: a case study of cardiovascular diseases and associated non-coding RNAs. Database (Oxford):2020
Yu B et al (2020) SubMito-XGBoost: predicting protein submitochondrial localization by fusing multiple feature information and eXtreme gradient boosting. Bioinformatics 36(4):1074–1081
Zhang B, Horvath S (2005) A general framework for weighted gene co-expression network analysis. Stat Appl Genet Mol Biol 4:17
Zhang J et al (2020) Accurate quantification of circular RNAs identifies extensive circular isoform switching events. Nat Commun 11(1):90
Zhang XO et al (2016) Diverse alternative back-splicing and alternative splicing landscape of circular RNAs. Genome Res 26(9):1277–1287
Zhao H et al (2020) LncTarD: a manually-curated database of experimentally-supported functional lncRNA-target regulations in human diseases. Nucleic Acids Res 48(D1):D118–d126
Zhao Z et al (2015) Co-LncRNA: investigating the lncRNA combinatorial effects in GO annotations and KEGG pathways based on human RNA-Seq data. Database (Oxford) 2015: 0910
Zheng Y et al (2019) Reconstruction of full-length circular RNAs enables isoform-level quantification. Genome Med 11(1):2
Zhou B et al (2018) EVLncRNAs: a manually curated database for long non-coding RNAs validated by low-throughput experiments. Nucleic Acids Res 46(D1):D100–d105
Zhou B et al (2021) EVLncRNAs 2.0: an updated database of manually curated functional long non-coding RNAs validated by low-throughput experiments. Nucleic Acids Res 49(D1):D86–d91

Machine Learning in Genomics

4

Sudipto Bhattacharjee, Abhirupa Ghosh, Banani Saha, and Sudipto Saha

Abstract

Machine learning (ML) techniques are implemented for handling and analyzing the large genomics datasets to address complex biological problems. It comprises computational algorithms that learn from data and identify unique patterns within the data. Analyzing genomics data using ML requires precise methods for representation, feature extraction, and feature selection that are specific to the data type, application, and algorithm. In this chapter, application of ML is discussed in genomics data analysis for sequence annotation, phenotype prediction, disease classification, precision medicine, prediction of epigenetic changes, protein-DNA interactions, and DNA-DNA interactions. In summary, ML is applied in different types of complex genomics data and is used in clinical applications and personalized medicine.

4.1 Introduction

Rapid development in the field of next-generation sequencing (NGS) has led to generation of large volumes of genomics data. Moreover, open-access databases such as Sequence Read Archive, ENSEMBL, and ENCODE have made it easier to store, access, and analyze these high-throughput datasets (Leinonen et al. 2010; Hunt et al. 2018; Davis et al. 2017). Genomics is represented as a broad sector of bioinformatics that includes approaches such as WGS, RNA-seq, DNA methylation,

S. Bhattacharjee · B. Saha
Department of Computer Science and Engineering, University of Calcutta, Kolkata, West Bengal, India

A. Ghosh · S. Saha (✉)
Division of Bioinformatics, Bose Institute, Kolkata, West Bengal, India

S. Singh (ed.), *Machine Learning and Systems Biology in Genomics and Health*, https://doi.org/10.1007/978-981-16-5993-5_4

ChIP-seq, Hi-C, and metagenomics. The downstream analyses of the large genomics datasets benefit in answering numerous biological questions associating genome (s) with disease diagnosis, progression, status, and therapeutics. Interpreting the large datasets requires implementation of artificial intelligence methods, like machine learning (ML) (Libbrecht and Noble 2015).

ML encompasses a set of algorithms that perform predictive tasks from training data by developing mathematical models that generalize the underlying training datasets as accurately as possible. ML algorithms can be broadly classified as supervised, unsupervised, and semi-supervised. In supervised learning, the training samples are labeled, and the objective of the trained model is to map the inputs to the labels, for example, classification and regression. In unsupervised learning, the training data is unlabeled, and the model tries to capture patterns within the data, for example, clustering. Semi-supervised learning falls between supervised and unsupervised learning that combines labeled and unlabeled data, for example, positive unlabeled (PU) learning. In recent years, deep learning (DL) has emerged as a popular ML approach in the fields of computer vision, pattern recognition, and natural language processing. The DL architecture consists of an artificial neural network (ANN) of multiple layers that can extract high-level features from a raw input. Hence, it is very useful to analyze genomic inputs like DNA sequences where classical feature extraction is hard.

The choice of a benchmark metric is very critical for evaluating the performance of ML models. For classification models, the metrics can be classified into threshold dependent (accuracy, sensitivity, and specificity) and independent (receiver operating characteristic). Accuracy is the proportion of correct predictions among the total number of predictions. Sensitivity measures the proportion of positives that are correctly identified, whereas specificity is the ratio of true negatives to all negatives. The receiver operating characteristic (ROC) is a plot of sensitivity with respect to (1-specificity) that is used for binary classifiers. The area under curve (AUC) of ROC plot is an aggregated evaluation of the model across all thresholds. For regression models, mean-squared error (MSE) and R^2-score are popular metrics. MSE is the average of the square of errors between the actual and predicted values. R^2-score is a measure of how well the regression predictions approximate the actual values.

ML and DL are being used for different applications such as identification of novel genomes, novel genes, or transcripts in genomes and variant prediction. ML and DL are also implemented in phenotype-genotype mapping, variant effect predictor, and integration of multi-omics data (Smith et al. 2020; Wang et al. 2020; Douglas et al. 2018). The genomic data analysis using ML can also solve clinical problems such as disease diagnosis, survival analysis, and drug response prediction (Yin et al. 2019a; Chaudhary et al. 2018; Stetson et al. 2014). This chapter discusses the prospects where ML has been implemented on different categories of genomics data as shown in Fig. 4.1. Different genomics data such as DNA sequence(s), transcriptomics data, epigenomics modifications, and DNA interactions were used as input features in ML and DL algorithms for efficient and accurate prediction of genomic characteristics, phenotypes, and diseases.

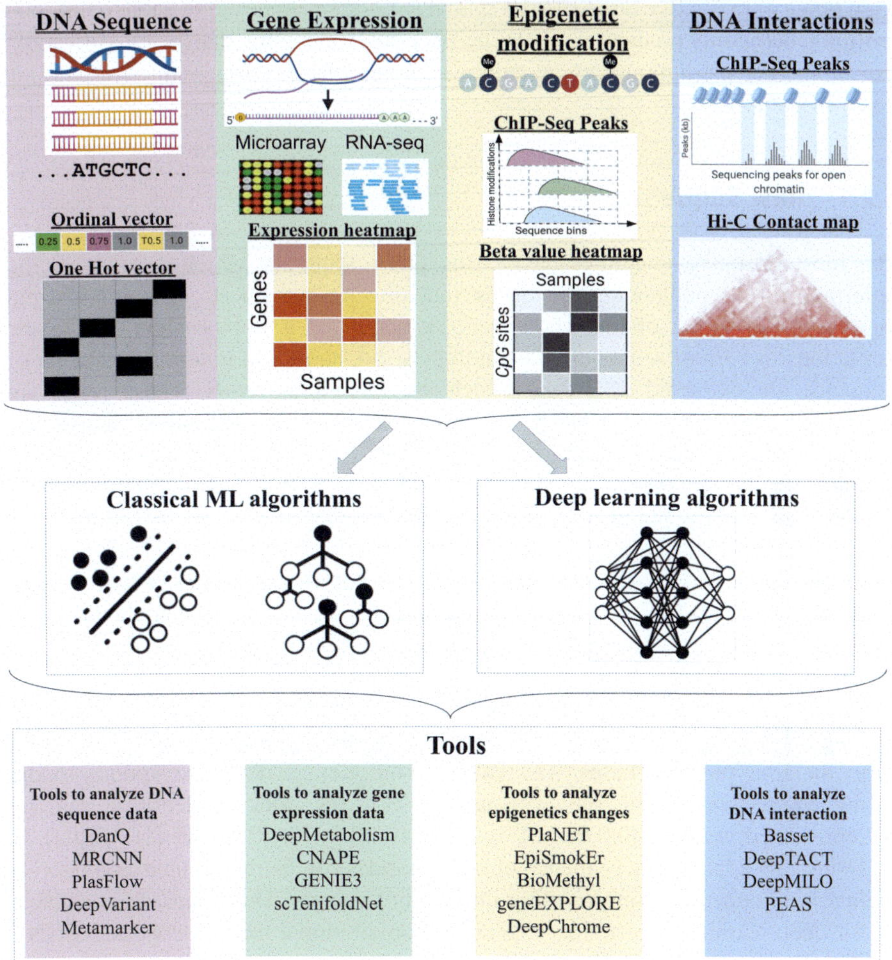

Fig. 4.1 Schematic diagram to showcase the role of ML in genomic data analysis

4.2 Analysis of Sequence Data

In recent years, the sequencing technology has expanded to massive parallel sequencing of large-scale genomics study. Downstream processing of genome sequence data provides insights of point or structural variations in the genome. Along with individual genomes, microbiome or environmental metagenomes are also explored using NGS technology. Machine learning is an important tool for analysis of raw as well as processed DNA sequence data. It is utilized to perform a wide range of predictive tasks such as spatial annotation, gene expression, chromatin profiling, taxonomy prediction, genotype-phenotype mapping, disease classification, and biomarker prediction. Few examples of tools of ML using sequence data are given in Table 4.1.

4.2.1 Raw Sequences

The most primitive genomics data is the DNA sequence which consists of the ordering of the nucleotides or bases (adenine-A, cytosine-C, guanine-G, and thymine-T). Reading of the whole genome sequence is not possible due to the limitations in current sequencing technologies. So, short and random fragments of sequence are extracted at a time which are known as *reads*. These reads are assembled according to overlaps in order to generate contiguous sequences, known as *contigs*, which are aligned with respect to a reference genome sequence. A mathematical representation of the sequence data is required for analysis using ML algorithms. Few popular approaches of such representation are as follows:

- *Ordinal encoding*: Each nucleotide in the sequence is mapped as ordinal values: A = 0.25, T = 0.5, G = 0.75 and C = 1.0. Thus, a sequence of n nucleotides can be represented as a vector $X = <x_1, x_2, \ldots, x_n>$ such that $x_i \in \{0.25, 0.5, 0.75, 1.0\}$.
- *One-hot encoding*: Every nucleotide in the sequence is mapped to a vector of length 4, known as one-hot vector. The four elements of such a vector correspond to the four possible bases. The value of the element that corresponds to the nucleotide is 1, and the rest are 0. So, the four nucleotides can be represented as: A = [0,0,0,1], T = [0,0,1,0], G = [0,1,0,0] and C = [1,0,0,0]. Using this approach, a sequence of n nucleotides can be represented as a $(n \times 4)$ matrix obtained by concatenating the one-hot vectors. This approach is suitable for deep learning algorithms, especially convolutional neural networks (CNN), that take matrices as input.
- *k-mer encoding*: In this approach, the sequence is first fragmented into subsequences of length k. For example, the sequence ATGCA can be fragmented into 2-mers as <AT, TG, GC, CA>. Then, each k-mer is encoded either using the ordinal or one-hot vector approach. The total number of elements in a dictionary of DNA sequence k-mers is 4^k as the number of bases is 4. So, a set of 4^k ordinal numbers or one-hot vectors of length 4^k are needed to represent each k-mer using the respective approaches. Zhou et al. (2011) represented bacterial DNA sequences as 2-mers (dinucleotides) and used their frequencies as features for support vector machine (SVM) to predict the species (Zhou et al. 2011). This approach consolidates the neighborhood information of a sequence.

To use sequence data as input features of ML algorithms, dimensionality reduction is often required. One option is to apply a hypothesis based on biological knowledge to obtain a region of interest (ROI) of the sequence that may provide

Table 4.1 ML-based tools to analyze sequence data

Sl. no.	Function	Tool	Input features	ML algorithm	Ref.
1.	Python application to predict chromatin profiles	DanQ	DNA sequence	CNN, BLSTM	Quang and Xie (2016)
2.	Prediction of gene expression	DeepExpression	DNA sequence	CNN, RNN	Zrimec et al. (2020)
3.	Python tool for prediction of DNA methylation levels	MRCNN	DNA sequence	CNN	Tian et al. (2019)
4.	Python tool to predict if a DNA sequence is of viral origin	ViraMiner	DNA sequence	CNN	Tampuu et al. (2019)
5.	Python tool for prediction of bacterial plasmid sequence and taxonomic classification	PlasFlow	DNA sequence	DNN	Krawczyk et al. (2018)
6.	Python software for classification between somatic and non-somatic SNVs	MutationSeq	VCF parameters	RF, BART, SVM, LgR	Ding et al. (2011)
7.	Somatic variant calling	NeoMutate; SNooPer; TNscope	BAM and VCF features; VCF parameters; BAM	LgR, NB, SVM, DT, RF, GB, neural network; RF; RF	Anzar et al. (2019); Spinella et al. (2016); Freed et al. (2018)
8.	Docker-based python tool for variant calling	DeepVariant	Images of read pileups	CNN	Poplin et al. (2018)
9.	Python tool for calling long deletions	DeepSV	Images of read pileups	CNN	Cai et al. (2019)
10.	R package for ensembled variant calling	BAYSIC; SMuRF	VCF files produced by different callers	Bayesian method; RF	Cantarel et al. (2014); Huang et al. (2019)
11.	Python tool for identification of biomarkers and prediction of colorectal cancer	MetaMarker	Metagenomic marker sequence	RF, SVM, neural network	Koohi-Moghadam et al. (2019)

(continued)

Table 4.1 (continued)

Sl. no.	Function	Tool	Input features	ML algorithm	Ref.
12.	Python tool for prediction of environmental and host phenotypes	MicroPheno	16S data	RF, SVM, DNN	Asgari et al. (2018)

Abbreviations: *BAM* binary alignment map, *BART* Bayesian additive regression tree, *BLSTM* bi-directional long short-term memory, *CNN* convolutional neural network, *DNN* deep neural network, *DT* decision tree, *GB* gradient boosting, *LgR* logistic regression, *LR* linear regression, *NB* Naive Bayes, *RF* random forest, *RNN* recurrent neural network, *SGD* stochastic gradient descent, *SNV* single nucleotide variant, *SVM* support vector machine, *VCF* variant call format.

sufficient features regarding the given problem. For example, Zeng et al. (2019) used a section of sequence data centered around the transcription start site (TSS) to select a promoter region as input (Zeng et al. 2019). In cases where such hypotheses cannot be applied, statistical methods are used, like principal component analysis (PCA) (Chen et al. 2019). While working with the one-hot vector approach, deep (nonlinear) feature selection using an auto-encoder is more suitable (Agarwal et al. 2019).

ML models are trained using raw sequences to predict different spatial annotations of DNA, like splice junctions (intron-exon, exon-intron, and neither) and promoter regions (Nguyen et al. 2016). DanQ and DeepSEA are deep learning tools that predict chromatin profiles: DNase sensitivity and histone modification markers (Quang and Xie 2016; Zhou and Troyanskaya 2015). The training data is preprocessed by splitting the genome into equal-sized bins. The sequences flanking the bins with at least one transcription factor (TF) binding event were selected as input. Both tools are similar in terms of training dataset and predicted chromatin factors, but DanQ uses simpler neural network architecture and outperforms DeepSEA. Prediction of gene expression and methylation levels directly from sequence data can reduce the extra effort and cost. DeepExpression computes gene expression using recurrent neural network (RNN) and CNN trained with DNA sequence data from combinations of regions: promoter, 5' UTR, 3' UTR, and terminator (Zrimec et al. 2020). MRCNN is a Python tool that uses DL to predict DNA methylation levels where training data contains 400 bp long one-hot encoded sequences centered around the methylation site in question (Tian et al. 2019). It achieved accuracy of >93% for sites from both *CpG* and non-*CpG* islands. ML is also used for bacterial and viral species classification and detection of plasmid sequences on the basis of raw sequence (Zhou et al. 2011; Tampuu et al. 2019; Krawczyk et al. 2018).

4.2.2 Variants

Genome of an organism contains numerous genetic mutations including single nucleotide polymorphisms (SNPs), insertions or deletions (InDels), and copy number variations (CNVs) leading to development of different genomic variants. The SNPs are further classified into synonymous, the mutations that are not reflected in amino acid sequence, and non-synonymous, the mutations that are translated into a different amino acid. Identifying or predicting variants is a major part in downstream analysis of genomics data. The quality filtered sequence reads are mapped to the reference genome and processed to remove duplicates and left aligning the InDels before passing onto the variant callers that determine any differences in sequences as variants over other artifacts. The variants are mapped to known phenotypes and annotated with their functional effect on the gene (Koboldt 2020). Most of the variant callers are based on Bayes' theorem that estimate the probability of a genotype being called as real polymorphism in the population given the observations of variants in the sequence reads and the prior probabilities of each genotype (Garrison and Marth 2012). Few others use heuristic approaches based on the parameters such as coverage, base quality, variant allele frequency, and statistical significance of the variants (Koboldt et al. 2013; Lai et al. 2016). The usual output of the variant calling step is Variant Call Format (VCF) file. The VCF file provides a range of information related to the quality of the read, quality of the variable nucleotide, mapping quality, allele frequency, and many more. In the diploid system, a site is termed heterozygous if the allele frequency of a non-reference allele is between 20% and 80%. However, in the case of non-diploid systems, multiple (not just two) alleles per site could be possible with allele frequencies between 0% and 100% (https://samtools.github.io/hts-specs/VCFv4.2.pdf).

Most of the variant callers use multiple of these features to call a variant. The conventional variant callers yield high false-positive predictions owing to their biggest challenge of distinguishing low-frequency variants with sequencing and alignment artifacts (Bian et al. 2018). The drawbacks of these callers are addressed using the ML approach to relatively improve the performance and accuracy of the variant calling process by treating it as a classification problem in ML. Few of the popular ML-based variant calling algorithms are SNooPer, MutationSeq, TNscope, and DeepVariant (Spinella et al. 2016; Ding et al. 2011; Freed et al. 2018; Poplin et al. 2018). MutationSeq integrates features of the mutations such as quality, coverage, location, strand bias, and others to develop classifiers by comparing random forest (RF), SVM, Bayesian additive regression tree (BART) and logistic regression models for somatic mutation prediction. Another somatic variant caller, SNooPer, is a random forest-based classifier trained using similar features from mpileup files (Spinella et al. 2016). VariFAST uses both BAM and VCF files to train a XGBoost model for high-quality variant detection (Zhang et al. 2019). DeepSV is a unique CNN-based tool for detecting long genomic deletions (Cai et al. 2019). Interestingly, few ML-based variant calling approaches integrate existing variant callers as an ensemble to obtain superior performance. Examples for this category are BAYSIC, SomaticSeq, SMuRF, and NeoMutate (Cantarel et al. 2014; Fang et al. 2015; Huang et al. 2019; Anzar et al. 2019). ML-based variant callers are further incorporated in pipelines for specific purposes such as dv-trio (Ip et al. 2020).

4.2.3 Metagenomics Data

The microbiome is observed to play a crucial role in human health, and perturbation of the normal host microbiome is often associated with disease conditions (Turnbaugh et al. 2007). Due to the importance of microbiome and evolution of sequencing technology, there is a latest boom of microbiome research that has also increased the amount of publicly available datasets. The microbiome studies mostly use either amplicon sequencing to identify the microbial composition or shotgun sequencing to identify both microbial composition and function. However, it is quite a challenge to handle and infer metagenomics data as it comprises multiple microbial genomes along with the host genome thus requiring sophisticated tools for efficient and accurate analysis of the microbiome data (Bharti and Grimm 2019).

The large amount of datasets has facilitated the use of ML algorithms to predict the microbial taxa, host phenotype, and microbial biomarkers. Metamarker is a de novo *k*-mer-based pipeline to predict metagenomic biomarkers and to level up multiple ML algorithms such as SVM and RF, and neural network models of colorectal cancer prediction (Koohi-Moghadam et al. 2019). Wu et al. (2018) have addressed multiclass classification problems for prediction of metagenomic biomarkers for multiple diseases through using logistic regression classifiers (Wu et al. 2018). MicroPheno is a pipeline to predict the environmental sites or disease phenotypes from shallow 16S data (Asgari et al. 2018). Uniquely it uses bootstrapping for proper selection of sampling size and *k*-mer size and uses ML algorithms such as RF, SVM, and deep neural network to predict the phenotypes.

4.3 Transcriptomics Data Analysis

Transcriptomics data contains the gene expression information. To measure the gene expression of a biological sample, DNA is converted to RNA fragments by transcription and intron splicing. Fragments of complementary DNA (cDNA) are synthesized from the RNA fragments with reverse transcription method. There are two popular techniques to determine gene expression levels from the cDNA: *RNA-seq* and *microarray analysis*. RNA-seq measures the expression of genes using NGS of cDNA (Wang et al. 2009). In microarray analysis, a DNA chip is used that contains a large number of biological probes to detect the expression level of genes represented by them (Govindarajan et al. 2012). The samples are fluorescently labeled and generate a signal upon binding with a probe.

The outputs of both microarray analysis and RNA-seq can be represented as a 2-D matrix. The columns of the matrix correspond to the samples, and rows correspond to the genes whose expression is measured. For RNA-seq, each element of the matrix contains the read count. The count matrix is normalized to eliminate experimental bias and to enable comparison of gene expression data across different experiments (Fundel et al. 2008). Some popular normalization methods are counts per million (CPM), transcript per kilobase million (TPM), and reads/fragments per kilobase of exon per million reads/fragments mapped (RPKM/FPKM). For microarray analysis, each element in the output matrix is the intensity of fluorescent signals emitted by the respective probe for each sample. A Python toolkit, SCANPY, is available for these preprocessing tasks and facilitates clustering and visualization of transcriptomics data (Wolf et al. 2018). It can also perform differential gene expression analysis and can be coupled with other Python-based ML packages.

One of the most popular applications of ML with transcriptomics data is genotype-to-phenotype mapping. A recent work by Smith et al. (2020) trained ML models with gene expression data to predict over 50 phenotypes that included binary and multiclass classification and survival analysis of various diseases (Smith et al. 2020). Both linear (PCA) and nonlinear (auto-encoder) dimensionality reduction were applied. They also proved that the method used for normalization can vary the performance of ML models. DeepMetabolism is an example of a DL model trained with gene expression data to predict phenotypes of *Escherichia coli* (Guo et al. 2017). The methodology is robust enough to be implemented for other organisms also. The CNAPE tool uses a logistic regression model trained with RNA-seq data to predict DNA copy number variation which is a critical marker for cancer diagnosis (Mu and Wang 2021). ML-based analysis of transcriptomics data is also performed for prediction of embryonic cell stages in humans with accuracy of 93% (AUC = 0.95) (Liang et al. 2020). ML can also be applied for gene regulatory network (GRN) construction using transcriptomics data. The R package, GENIE3, uses an ensemble of regression models for the construction of GRN from gene expression matrix (Huynh-Thu et al. 2010). Each model attempts to predict the expression of a gene p using the expression of all other genes as explanatory variables such that the respective regression model weights represent the edge-weights in the GRN directed to p. Another R package, scTenifoldNet, can construct GRN from transcriptomics data in an unsupervised manner (Osorio et al. 2020). Few examples of tools of ML using transcriptomics data are given in Table 4.2.

4.4 Analysis of Epigenetic Modification Data

DNA methylation and histone modifications are two primary epigenetic processes. Perturbations in either of these processes impact gene expression levels, disease mechanism, and other biological processes. NGS has equipped the epigenetic studies on a global scale across the genome using methods such as bisulfite-seq and ChIP-seq. Additionally, application of ML facilitates efficient analysis of these data. Few ML tools available for prediction of epigenetic modification are listed in Table 4.3.

4.4.1 DNA Methylation Data

DNA methylation (DNAm) is a process which results in addition of methyl groups (CH_3) to the DNA, generally at cytosine (C) bases mainly in *CpG* sites—the "*5'–C–phosphate–G–3'*" sequence of nucleotides. Techniques such as bisulfite sequencing (BS-seq) and BeadChip Array analysis are used to detect methylation levels. For

Table 4.2 ML-based tools trained with expression data

Sl. no.	Function	Tool	Input features	ML algorithm	Ref.
1.	Python tool phenotype prediction of *E. coli* (a) Growth rate (h^{-1}) (b) Succinate yield (cmol/cmol) (c) Ethanol yield (cmol/cmol)	DeepMetabolism	Gene expression	DNN	Guo et al. (2017)
2.	R-based command-line tool to predict DNA copy number variation	CNAPE	RNA-seq	LgR with LASSO	Mu and Wang (2021)
3.	R package for 1. Construction of gene regulatory network 2. Ranking of regulatory links	GENIE3	Microarray	RF, ET	Huynh-Thu et al. (2010)
4.	R package for construction of gene regulatory network	scTenifoldNet	RNA-seq	PCR	Osorio et al. (2020)
5.	Web application for prediction of human embryonic cell development stages	EmPredictor	RNA-seq	SVM	Liang et al. (2020)

Abbreviations: *DNN* deep neural network, *ET* extra tree, *LASSO* least absolute shrinkage and selection, *LgR* logistic regression, *PCR* principal component regression, *RF* random forest, *SVM* support vector machine.

each such *CpG* site, DNAm data is generally represented as beta values or M-values (Du et al. 2010). Other information such as the location of those *CpG* sites with respect to the genome and TSS are also present. The distance of the *CpG* site from the TSS can be used to annotate if it is a promoter region. Methylation and gene expression are highly correlated because methylation in the promoter region leads to suppression of gene expression. Hence, the *CpG* sites centered around the TSS are significant for ML applications.

The DNAm data corresponding to a sequence of *CpG* sites $<c_1, c_2, \ldots, c_n>$ can be represented as a vector $M = [x_1, x_2, \ldots, x_n]$. The elements, x_i, of the vector can either be the methylation state *{0, 1}* or the quantification of methylation (beta values or M-values) at the corresponding location c_i. DNAm data is known to be high dimensional with low sample size due to the abundance of *CpG* sites in the genome. Hence, for practical applications such as disease-normal classification and disease progression, severity, and recurrence prediction, the methylation data from *CpG* sites of disease-specific genes is used with TSS as the reference point in the genome. PyMethylProcess is a command-line tool that contains workflows for preprocessing DNAm data before analysis using ML algorithms (Levy et al. 2019). It can download datasets from public databases and perform feature selection, quality control, and normalization and store results as a data structure named MethylationArray.

Table 4.3 Applications of ML using different modification data

Sl. no.	Function	Tool	Input features	ML algorithm	Ref.
1.	R package for ethnicity prediction from placental samples	PlaNET	BeadChip	LgR, KNN, SVM	Yuan et al. (2019)
2.	R package for prediction of smoking status	EpiSmokEr	BeadChip	LASSO	Bollepalli et al. (2019)
3.	R tool for prediction of gene expression	BioMethyl; geneEXPLORE	BeadChip	LR; elastic net	Wang et al. (2019); Kim et al. (2020)
4.	Perl framework for prediction of gene expression	MethylXcan	BeadChip	LR, LASSO	Zhong et al. (2019)
5.	Prediction of gene expression	DeepChrome	ChIP-seq [a]	CNN	Singh et al. (2016)
6.	Prediction of differential gene expression	DeepDiff	ChIP-seq [a]	LSTM	Sekhon et al. (2018)

Abbreviations: *KNN* k-Nearest Neighbor, *LASSO* least absolute shrinkage and selection operator, *LSTM* long short-term memory, *LgR* logistic regression, *LR* linear regression, *RF* random forest, *SVM* support vector machine.
[a]ChIP-Seq was used for profiling histone modification (methylation and acetylation).

MethylNet is a command-line framework for analyzing DNAm datasets using deep learning (Levy et al. 2020). It provides functionalities for feature selection using auto-encoders, hyperparameter optimization, classification, regression, and interpretation of predictions. It can process the MethylationArray datasets generated by PyMethylProcess. It also contains trained models for prediction of age, smoking status, and different cancer types.

Methylation of *CpG* sites is known to inhibit gene expression. BioMethyl is an R package that uses linear regression models to exploit this correlation and predict gene expression from DNAm data of 37 cancer types (Wang et al. 2019). MethylXcan used DNAm data of non-cancer tissue to develop least absolute shrinkage and selection operator (LASSO) penalized regression models to predict gene expression (Zhong et al. 2019). geneEXPLORE showed that using long-range methylations of $\geq$10 Mb can improve the performance of gene expression predicting models (Kim et al. 2020). Analysis of DNAm patterns is also highly beneficial in forensic studies due to its correlation with cellular aging (Horvath et al. 2012). A number of ML models have been developed leveraging this correlation to form an epigenetic clock and predict age using various algorithms (Vidaki et al. 2017; Aliferi et al. 2018; Thong et al. 2021). ML is also used in population stratification tasks on the basis of methylation data (Barfield et al. 2014). PlaNET is an R package that uses placental DNAm data to develop various ML models for ethnicity classification with 94% accuracy (Yuan et al. 2019). ML is also used for classification of smokers and prediction of daily consumption using methylation data (Philibert et al. 2020).

EpiSmokEr is an R package for classifying smokers as "never smoker", "former smoker," and "current smoker" (Bollepalli et al. 2019).

4.4.2 Histone Modifications

Posttranslational modifications of histone proteins such as methylation, acetylation, phosphorylation, and others together are termed as histone modifications. Histone modification plays a crucial role in chromosomal conformation and gene expression (Kouzarides 2007). ChIP-seq (Chromatin ImmunoPrecipitation followed by sequencing) is a NGS technique to determine the specific DNA-binding protein or epigenetic markers including histone modification (Park 2009). The epigenetic information obtained from ChIP-seq data are often used to train machine learning models for prediction of gene expression levels. Karlić et al. (2010) have shown that linear regression model can infer connection of 19 lysine (K) and arginine (N) methylation of histone and three promoter site modifications to gene expression (Karlić et al. 2010). Similarly, Dong et al. (2012) uses 11 histone modifications and one histone variant to develop a two-step prediction model consisting of random forest classification model and a regression model (linear, random forest, and MARS regressors) that shows correlation between chromatin features and expression levels across different cell types (Dong et al. 2012). Cheng et al. (2011) have applied a SVM-based regression model learned from histone modifications of *C. elegans* and predicted gene expressions in various species including humans (Cheng et al. 2011). Motlagh et al. (2020) have used only histone acetylation marker information to train a multilayer perceptron neural network for the prediction of specific transcription factors' binding sites (Banirazi Motlagh et al. 2020). DeepChrome is a CNN framework that learns histone modification markers and their interactions to predict gene expression (Singh et al. 2016). DeepDiff is an DL-based tool for prediction of differential gene expression between two cell types using histone modifications (Sekhon et al. 2018). Taken together, several ML-based tools are available to analyze DNA methylation and histone modification data.

4.5 DNA Interactions

The advancements in NGS technologies have led the path for improved techniques to explore DNA accessibility and nucleosome positioning such as DNase-seq, FAIRE-seq, MNase-seq, and ATAC-seq. Along with histone modification profiling, ChIP-seq is also used for characterization of transcription factors and histone interacting sites. Hi-C is another high-throughput method that detects all possible DNA-DNA interactions in the genome and captures the 3D genome organization. ChIA-PET combines Hi-C and ChIP-seq for the genome-wide prediction of interaction mediated by a transcription factor of interest.

Machine learning is explored in predicting the specific protein binding sequence (s) in DNA or RNA, enhancer site, regulatory regions, methylation state, and gene

Table 4.4 ML-based tools trained with different interaction data

Sl. no.	Function	Tool	Input features	ML algorithm	Ref.
1.	Python-based command-line tool for prediction of DNA accessibility	Basset	DNase-seq	CNN	Kelley et al. (2016)
2.	Python tool for prediction of histone modification sites	DeepHistone	DNase-seq	CNN	Yin et al. (2019b)
3.	Python tool for prediction of chromatin contacts	DeepTACT	DNase-seq	CNN and RNN with bootstrap	Li et al. (2019)
4.	Predictions of enhancers	PEAS	ATAC-seq	DNN	Thibodeau et al. (2018)
5.	Prediction of variant effect on insulator loops	DeepMILO	ChIA-PET	CNN and RNN	Trieu et al. (2020)

Abbreviations: *CNN* convolutional neural network, *DNN* deep neural network, *RNN* recurrent neural network.

expression using chromatin accessible data. Basset is a deep CNN-based tool that predicts DNA accessibility using DNase-seq peak BED format files with high level of precision (Kelley et al. 2016). It also assigns cell-type-specific scores to variants from GWAS to predict accessibility changes due to the mutations. DeepHistone is another deep learning-based method that uses sequence information and DNase-seq data to train a CNN model for the prediction of histone modification sites in a given sequence (Yin et al. 2019b). Another DL-based method, DeepTACT, uses DNase-seq data along with sequence information for training a bootstrapping deep learning model for the prediction of contacts between regulatory elements (Li et al. 2019). Interestingly, it can use promoter capture HiC data to predict promoter-enhancer as well as promoter-promoter contacts. Similar to previously mentioned methods, the predicted contact information from DeepTACT can be integrated with the GWAS data to identify disease associated markers. PEAS is a neural network-based tool that uses ATAC-seq data for the prediction of enhancers across cell types (Thibodeau et al. 2018). Farré et al. (2018) have applied CNN to predict chromatin conformation from sequence data using sequence of bound proteins from ChIP-seq as input data and scored them using transcriptomics data and compared with chromatin structural information from Hi-C contact maps for validation (Farré et al. 2018). They also addressed the inverse problem of predicting the 1D chromatin sequence from local Hi-C contact maps using another independent DNN. To improve their primary model, they have incorporated the 1D chromatin sequence from the second model into the first. DeepMILO is a CNN- and RNN-based model trained, validated, and tested on CTCF and Cohesin ChIA-PET data to predict the effect of mutations on the CTCF/Cohesin insulator loops (Trieu et al. 2020). Few ML tools using DNA interaction data are listed in Table 4.4.

4.6 Integrating Multi-Omics Data for Analysis

Integrated analysis of different NGS approaches such as WGS, transcriptomics, epigenomics, and metagenomics with other "OMICS" data gives a better understanding of biological mechanisms compared to any single-layer analysis. However, integrating heterogeneous datatypes poses challenges and requires human interventions throughout. To reduce the tedious and time-consuming manual labor, ML has been implemented in many cases. One of the methods to reduce the heterogeneity issue is MOFA, a Bayesian model-based tool to decompose the sources for identifying the major variations among data. It is implemented in both R and Python. An R package, integrOmics, implements canonical correlation analysis (CCA) and partial least squares (PLS) regression to integrate any two types of omics data (such as transcriptomics, proteomics, metabolomics, and multiple datasets from different technology platforms) of same samples (Lê Cao et al. 2009). Multi-view Factorization AutoEncoder (MAE) is a ML-based method that integrates known biological network information from databases like STRING or REACTOME with gene expression, miRNA expression, DNA methylation, or/and other omics data to reduce the overfitting problem in case of small sample size and large feature dimensionality.

There are numbers of methods using different ML algorithms on multi-omics data to answer very specific biological questions such as finding GRN, clinical outcome or disease prediction, and survival analysis. iOmicsPass applies a modified nearest shrunken centroid (NSC) algorithm on mRNA expression, protein expression, and copy number variation data along with the user-provided network data to predict gene subnetwork for specific phenotypes (Koh et al. 2019). Douglas et al. (2018) have integrated human genome variant data with 16S and shotgun metagenomics data from pediatric Crohn's disease and normal samples to classify disease state and treatment response using separate RF models (Douglas et al. 2018). RF and LASSO logistic regression models were used by Yachida et al. (2019) on WGS shotgun metagenomics and microbial metabolomics of colorectal cancer samples to predict metagenomic and metabolic markers that help in differentiating colorectal cancer from controls (Yachida et al. 2019). Hollister et al. (2019) have combined metagenomic and metabolic markers to feed a random forest classifier to differentiate pediatric inflammatory bowel syndrome (IBS) patients from healthy controls (Hollister et al. 2019). Chaudhary et al. (2018) identified differential survival of hepatocellular carcinoma (HCC) patients by using a multi-ML framework comprising auto-encoder, k-means clustering, and SVM on integrated RNA-seq, MiRNA-seq, and DNA methylation data (Chaudhary et al. 2018). In conclusion, genomic features were integrated with other omics features using ML-based techniques.

4.7 Clinical Applications of ML Using Genomics Data

The prime objective of genomics data analysis using ML is to answer the clinical questions to provide better therapeutics. The spectrum of clinical problems is quite large including disease diagnosis and prediction of risk, progression, and survival. Identification of disease-specific genomic causality such as risk genes and variants is also a major challenge. ML can provide a better understanding of the biological mechanisms of cancers, genetic disorders, metabolic, and infectious diseases (Hamzeh et al. 2020; Park et al. 2020; López et al. 2018). In recent years, the concept of precision therapeutics is in the limelight and multi-omics data analyses plays the pivotal role in predicting personalized medicine for specific patients or stratify patients into treatment groups. The computational models for prediction of drug response paved the path for personalized medicine prediction. Few examples of clinical applications of ML using genomic data are given in Table 4.5.

4.8 Drawbacks and Challenges

There are few challenges in implementing ML in genomic datasets. The number of features in genomics data is generally higher than the number of samples. Training ML models on such data can cause *overfitting* where the models are trained too exactly and unable to generalize the data. It reduces the model's ability to predict unknown samples. One option to overcome overfitting is to use feature selection to remove insignificant features. The other option includes the use of ensemble methods and cross validation. Algorithm-specific techniques such as dropout of neurons in neural networks and penalization in regression algorithms can also be used. The large feature size can also make the genomic datasets to be huge, even with small sample size. This increases the *computational cost*, both in terms of space and time complexity. ML models deployed on GPUs use parallelism to make the training process faster. Cloud computing is another option to deal with the computational burden. While using genomic data, the biological causality inferred by the ML models is often more important than the prediction accuracies. The ML algorithms, especially deep learning, are referred to as "black box" because the *detection of inferences* between the data and the predicted outcome can be quite difficult. Interpretation of ML models is an open problem. An approach can be to vary a set of input features and check the variation of output. For CNN, analysis of a heatmap of the convolutional filter may provide an insight into the learning of localized features. In tasks where the ML model identifies significant genes or SNPs, the prediction output effectively provides hypotheses which require further *experimental validation*. For classification problems such as tumor-normal discrimination, samples tend to be *imbalanced* that makes the learning biased. A straightforward solution is to randomly down sample the majority class to balance the dataset but with the risk of losing information. Also, if the size of the minority class is too low, this is not a suitable approach. A more practical solution would be to calculate the sensitivity and specificity of the prediction instead of accuracy to evaluate the

Table 4.5 Few clinical applications of ML using genomic data

Sl. no.	Disease	Objective	Input features	ML algorithm	Ref.
1.	Obesity	1. Regression on body mass index 2. Classification of weight status	GE, SNP	1. LR 2. SVM	Joseph et al. (2018)
2.	Type 2 diabetes	Disease prediction and prediction of significant SNPs	SNP; SNP, DNAm	RF; LgR, NB, SVM, CART	López et al. (2018); Hathaway et al. (2019)
3.	Lung cancer; breast cancer; prostate cancer	Detection of subtypes; prediction of recurrence; prediction of tumor location and identification of differentially expressed genes	GE; DNAm; GE	SVM; DT, GB, RF, NB, KNN, AdaBoost; NB, SVM, RF	Yuan et al. (2020); Macías-García et al. (2020); Hamzeh et al. (2020)
4.	Cancer	Classification of different cancer types	DNAm; ATAC-seq	DNN	Zheng and Xu (2020); Yang et al. (2020)
5.	COVID-19	Disease prediction	SNP	RF, DNN	Wang et al. (2020)
6.	Asthma	1. Identification of disease causing SNPs 2. Disease prediction	SNP	KNN, SVM	Gaudillo et al. (2019)
7.	Amyotrophic lateral sclerosis (ALS); Alzheimer's disease	Disease prediction	SNP; GE, DNAm	CNN; DNN	Yin et al. (2019a); Park et al. (2020)
8.	Huntington's disease; inflammatory bowel disease	Identification of associated genes	GE	DT, RF, GLM; RF, SVM, EGB, EN	Cheng et al. (2020); Isakov et al. (2017)
9.	Cancer	Drug response prediction	CNV, GE, DNAm; GE, CNV, mutation	RF, SVM, elastic net; RF	Wan and Pal (2014); Stetson et al. (2014)

Abbreviations: *CART* classification and regression tree, *CNN* convolutional neural network, *DNN* deep neural network, *DNAm* DNA methylation, *DT* decision tree, *EGB* extreme gradient boosting, *EN* elastic net, *GB* gradient boosting, *GE* gene expression, *GLM* generalized linear model, *KNN*, k-nearest neighbor, *LgR* logistic regression, *LR* linear regression, *NB* Naive Bayes, *RF* random forest, *SNP* single nucleotide polymorphism, *SVM* support vector machine.

model performance. Another drawback is that the genomic datasets used for training models are generally imbalanced in terms of population ethnicity. This hinders the models' ability to accommodate any *population variation* in the prediction tasks.

4.9 Future Scope

The use of ML tools and methods has considerably increased for genomics data analysis with the ultimate aim of improved clinical decision-making. Continuous efforts are still made to enhance the efficiency of such methods. To that end, the field of radiogenomics has recently emerged that attempts to find associations between attributes of genomics and radiological investigations such as high-resolution computed tomography (HRCT) and positron-emission tomography (PET) (Lo Gullo et al. 2020). Fonseca et al. revealed that identification of genomic associations with pulmonary function tests (PFT) can be helpful in diagnosis of lung diseases (Fonseca et al. 2020). Similar associations with electrocardiogram (ECG) features were also proven important for heart disease prediction (Ritchie et al. 2013). There is an opportunity for researchers to integrate features of genomics and clinical investigation data and apply ML algorithms to develop automated tools that can aid clinicians in making an informed decision.

References

Agarwal V, Reddy NJK, Anand A (2019) Unsupervised representation learning of DNA sequences

Aliferi A, Ballard D, Gallidabino MD, Thurtle H, Barron L, Syndercombe Court D (2018) DNA methylation-based age prediction using massively parallel sequencing data and multiple machine learning models. Forensic Sci Int Genet 37:215–226. https://doi.org/10.1016/j.fsigen.2018.09.003

Anzar I, Sverchkova A, Stratford R, Clancy T (2019) NeoMutate: an ensemble machine learning framework for the prediction of somatic mutations in cancer. BMC Med Genomics 12(1):63. https://doi.org/10.1186/s12920-019-0508-5

Asgari E, Garakani K, McHardy AC, Mofrad MRK (2018) MicroPheno: predicting environments and host phenotypes from 16S rRNA gene sequencing using a k-mer based representation of shallow sub-samples. Bioinformatics 34(13):i32–i42. https://doi.org/10.1093/bioinformatics/bty296

Banirazi Motlagh N, Mohammadpour Esfahani B, Ashrafi B, Zare-Mirakabad F (2020) The assessment of histone acetylation marks in the vicinity of transcription factor binding sites in human CD4 + T cells using information theory methods. Comput Biol Chem 86:107232. https://doi.org/10.1016/j.compbiolchem.2020.107232

Barfield RT et al (2014) Accounting for population stratification in DNA methylation studies. Genet Epidemiol 38(3):231–241. https://doi.org/10.1002/gepi.21789

Bharti R, Grimm DG (2019) Current challenges and best-practice protocols for microbiome analysis. Brief Bioinform 22(1):178–193. https://doi.org/10.1093/bib/bbz155

Bian X et al (2018) Comparing the performance of selected variant callers using synthetic data and genome segmentation. BMC Bioinformatics 19(1):429. https://doi.org/10.1186/s12859-018-2440-7

Bollepalli S, Korhonen T, Kaprio J, Anders S, Ollikainen M (2019) EpiSmokEr: a robust classifier to determine smoking status from DNA methylation data. Epigenomics 11(13):1469–1486. https://doi.org/10.2217/epi-2019-0206

Cai L, Wu Y, Gao J (2019) DeepSV: accurate calling of genomic deletions from high-throughput sequencing data using deep convolutional neural network. BMC Bioinformatics 20(1):665. https://doi.org/10.1186/s12859-019-3299-y

Cantarel BL, Weaver D, McNeill N, Zhang J, Mackey AJ, Reese J (2014) BAYSIC: a Bayesian method for combining sets of genome variants with improved specificity and sensitivity. BMC Bioinformatics 15(1):104. https://doi.org/10.1186/1471-2105-15-104

Chaudhary K, Poirion OB, Lu L, Garmire LX (2018) Deep learning{\textendash}based multi-omics integration robustly predicts survival in liver cancer. Clin Cancer Res 24(6):1248–1259. https://doi.org/10.1158/1078-0432.CCR-17-0853

Chen Z et al (2019) iLearn: an integrated platform and meta-learner for feature engineering, machine-learning analysis and modeling of DNA, RNA and protein sequence data. Brief Bioinform 21(3):1047–1057. https://doi.org/10.1093/bib/bbz041

Cheng C et al (2011) A statistical framework for modeling gene expression using chromatin features and application to modENCODE datasets. Genome Biol 12(2):R15. https://doi.org/10.1186/gb-2011-12-2-r15

Cheng J, Liu H-P, Lin W-Y, Tsai F-J (2020) Identification of contributing genes of Huntington's disease by machine learning. BMC Med Genomics 13(1):176. https://doi.org/10.1186/s12920-020-00822-w

Davis CA et al (2017) The encyclopedia of DNA elements (ENCODE): data portal update. Nucleic Acids Res 46(D1):D794–D801. https://doi.org/10.1093/nar/gkx1081

Ding J et al (2011) Feature-based classifiers for somatic mutation detection in tumour–normal paired sequencing data. Bioinformatics 28(2):167–175. https://doi.org/10.1093/bioinformatics/btr629

Dong X et al (2012) Modeling gene expression using chromatin features in various cellular contexts. Genome Biol 13(9):R53. https://doi.org/10.1186/gb-2012-13-9-r53

Douglas GM et al (2018) Multi-omics differentially classify disease state and treatment outcome in pediatric Crohn's disease. Microbiome 6(1):13. https://doi.org/10.1186/s40168-018-0398-3

Du P et al (2010) Comparison of Beta-value and M-value methods for quantifying methylation levels by microarray analysis. BMC Bioinformatics 11(1):587. https://doi.org/10.1186/1471-2105-11-587

Fang LT et al (2015) An ensemble approach to accurately detect somatic mutations using SomaticSeq. Genome Biol 16(1):197. https://doi.org/10.1186/s13059-015-0758-2

Farré P, Heurteau A, Cuvier O, Emberly E (2018) Dense neural networks for predicting chromatin conformation. BMC Bioinformatics 19(1):372. https://doi.org/10.1186/s12859-018-2286-z

Fonseca H et al (2020) Genomic regions 10q22.2, 17q21.31, and 2p23.1 can contribute to a lower lung function in African descent populations. Genes (Basel) 11:9. https://doi.org/10.3390/genes11091047

Freed D, Pan R, Aldana R (2018) TNscope: accurate detection of somatic mutations with haplotype-based variant candidate detection and machine learning filtering. bioRxiv. https://doi.org/10.1101/250647

Fundel K, Haag J, Gebhard PM, Zimmer R, Aigner T (2008) Normalization strategies for mRNA expression data in cartilage research. Osteoarthr Cartil 16(8):947–955. https://doi.org/10.1016/j.joca.2007.12.007

E. Garrison and G. Marth, "Haplotype-based variant detection from short-read sequencing." 2012

Gaudillo J et al (2019) Machine learning approach to single nucleotide polymorphism-based asthma prediction. PLoS One 14(12):e0225574. https://doi.org/10.1371/journal.pone.0225574

Govindarajan R, Duraiyan J, Kaliyappan K, Palanisamy M (2012) Microarray and its applications. J Pharm Bioallied Sci 4(6):310–312. https://doi.org/10.4103/0975-7406.100283

Guo W, Xu YE, Feng X (2017) DeepMetabolism: a deep learning system to predict phenotype from genome sequencing bioRxiv, https://doi.org/10.1101/135574

Hamzeh O, Alkhateeb A, Zheng J, Kandalam S, Rueda L (2020) Prediction of tumor location in prostate cancer tissue using a machine learning system on gene expression data. BMC Bioinformatics 21(2):78. https://doi.org/10.1186/s12859-020-3345-9

Hathaway QA et al (2019) Machine-learning to stratify diabetic patients using novel cardiac biomarkers and integrative genomics. Cardiovasc Diabetol 18(1):78. https://doi.org/10.1186/s12933-019-0879-0

Hollister EB et al (2019) Leveraging human microbiome features to diagnose and stratify children with irritable bowel syndrome. J Mol Diagn 21(3):449–461. https://doi.org/10.1016/j.jmoldx.2019.01.006

Horvath S et al (2012) Aging effects on DNA methylation modules in human brain and blood tissue. Genome Biol 13(10):R97. https://doi.org/10.1186/gb-2012-13-10-r97

Huang W, Guo YA, Muthukumar K, Baruah P, Chang MM, Jacobsen Skanderup A (2019) SMuRF: portable and accurate ensemble prediction of somatic mutations. Bioinformatics 35(17):3157–3159. https://doi.org/10.1093/bioinformatics/btz018

Hunt SE et al (2018) Ensembl variation resources. Database 2018:119. https://doi.org/10.1093/database/bay119

Huynh-Thu VA, Irrthum A, Wehenkel L, Geurts P (2010) Inferring regulatory networks from Expression data using tree-based methods. PLoS One 5(9):e12776. https://doi.org/10.1371/journal.pone.0012776

Ip EKK, Hadinata C, Ho JWK, Giannoulatou E (2020) dv-trio: a family-based variant calling pipeline using DeepVariant. Bioinformatics 36(11):3549–3551. https://doi.org/10.1093/bioinformatics/btaa116

Isakov O, Dotan I, Ben-Shachar S (2017) Machine learning–based gene prioritization identifies novel candidate risk genes for inflammatory bowel disease. Inflamm Bowel Dis 23(9):1516–1523. https://doi.org/10.1097/MIB.0000000000001222

Joseph PV, Wang Y, Fourie NH, Henderson WA (2018) A computational framework for predicting obesity risk based on optimizing and integrating genetic risk score and gene expression profiles. PLoS One 13(5):e0197843. https://doi.org/10.1371/journal.pone.0197843

Karlić R, Chung H-R, Lasserre J, Vlahovicek K, Vingron M (2010) Histone modification levels are predictive for gene expression. Proc Natl Acad Sci U S A 107(7):2926–2931. https://doi.org/10.1073/pnas.0909344107

Kelley DR, Snoek J, Rinn JL (2016) Basset: learning the regulatory code of the accessible genome with deep convolutional neural networks. Genome Res 26(7):990–999. https://doi.org/10.1101/gr.200535.115

Kim S, Park HJ, Cui X, Zhi D (2020) Collective effects of long-range DNA methylations predict gene expressions and estimate phenotypes in cancer. Sci Rep 10(1):3920. https://doi.org/10.1038/s41598-020-60845-2

Koboldt DC (2020) Best practices for variant calling in clinical sequencing. Genome Med 12(1):91. https://doi.org/10.1186/s13073-020-00791-w

Koboldt DC, Larson DE, Wilson RK (2013) Using VarScan 2 for germline variant calling and somatic mutation detection. Curr Protoc Bioinformatics 44(1):15.4.1–15.4.17. https://doi.org/10.1002/0471250953.bi1504s44

Koh HWL, Fermin D, Vogel C, Choi KP, Ewing RM, Choi H (2019) iOmicsPASS: network-based integration of multiomics data for predictive subnetwork discovery. Syst Biol Appl 5(1):22. https://doi.org/10.1038/s41540-019-0099-y

Koohi-Moghadam M et al (2019) MetaMarker: a pipeline for de novo discovery of novel metagenomic biomarkers. Bioinformatics 35(19):3812–3814. https://doi.org/10.1093/bioinformatics/btz123

Kouzarides T (2007) Chromatin modifications and their function. Cell 128(4):693–705. https://doi.org/10.1016/j.cell.2007.02.005

Krawczyk PS, Lipinski L, Dziembowski A (2018) PlasFlow: predicting plasmid sequences in metagenomic data using genome signatures. Nucleic Acids Res 46(6):e35–e35. https://doi.org/10.1093/nar/gkx1321

Lai Z et al (2016) VarDict: a novel and versatile variant caller for next-generation sequencing in cancer research. Nucleic Acids Res 44(11):e108–e108. https://doi.org/10.1093/nar/gkw227

Lê Cao K-A, González I, Déjean S (2009) integrOmics: an R package to unravel relationships between two omics datasets. Bioinformatics 25(21):2855–2856. https://doi.org/10.1093/bioinformatics/btp515

Leinonen R, Sugawara H, Martin NSDCS (2010) The sequence read archive. Nucleic Acids Res 39: D19–D21. https://doi.org/10.1093/nar/gkq1019

Levy JJ, Titus AJ, Petersen CL, Chen Y, Salas LA, Christensen BC (2020) MethylNet: an automated and modular deep learning approach for DNA methylation analysis. BMC Bioinformatics 21(1):108. https://doi.org/10.1186/s12859-020-3443-8

Levy JJ, Titus AJ, Salas LA, Christensen BC (2019) PyMethylProcess—convenient high-throughput preprocessing workflow for DNA methylation data. Bioinformatics 35(24):5379–5381. https://doi.org/10.1093/bioinformatics/btz594

Li W, Wong WH, Jiang R (2019) DeepTACT: predicting 3D chromatin contacts via bootstrapping deep learning. Nucleic Acids Res 47(10):e60–e60. https://doi.org/10.1093/nar/gkz167

Liang P et al (2020) Machine learning of single-cell transcriptome highly identifies mRNA signature by comparing F-score selection with DGE analysis. Mol Ther - Nucleic Acids 20: 155–163. https://doi.org/10.1016/j.omtn.2020.02.004

Libbrecht MW, Noble WS (2015) Machine learning applications in genetics and genomics. Nat Rev Genet 16(6):321–332. https://doi.org/10.1038/nrg3920

Lo Gullo R, Daimiel I, Morris EA, Pinker K (2020) Combining molecular and imaging metrics in cancer: radiogenomics. Insights Imaging 11(1):1. https://doi.org/10.1186/s13244-019-0795-6

López B, Torrent-Fontbona F, Viñas R, Fernández-Real JM (2018) Single nucleotide polymorphism relevance learning with random forests for type 2 diabetes risk prediction. Artif Intell Med 85:43–49. https://doi.org/10.1016/j.artmed.2017.09.005

Macías-García L, Martínez-Ballesteros M, Luna-Romera JM, García-Heredia JM, García-Gutiérrez J, Riquelme-Santos JC (2020) Autoencoded DNA methylation data to predict breast cancer recurrence: machine learning models and gene-weight significance. Artif Intell Med 110: 101976. https://doi.org/10.1016/j.artmed.2020.101976

Mu Q, Wang J (2021) CNAPE: a machine learning method for copy number alteration prediction from gene expression. IEEE/ACM Trans Comput Biol Bioinforma 18(1):306–311. https://doi.org/10.1109/TCBB.2019.2944827

Nguyen N et al (2016) DNA sequence classification by convolutional neural network. J Biomed Sci Eng 9(5):280–286. https://doi.org/10.4236/jbise.2016.95021

Osorio D, Zhong Y, Li G, Huang JZ, Cai JJ (2020) scTenifoldNet: a machine learning workflow for constructing and comparing transcriptome-wide gene regulatory networks from single-cell data. Patterns 1(9):100139. https://doi.org/10.1016/j.patter.2020.100139

Park C, Ha J, Park S (2020) Prediction of Alzheimer's disease based on deep neural network by integrating gene expression and DNA methylation dataset. Expert Syst Appl 140:112873. https://doi.org/10.1016/j.eswa.2019.112873

Park PJ (2009) ChIP–seq: advantages and challenges of a maturing technology. Nat Rev Genet 10(10):669–680. https://doi.org/10.1038/nrg2641

Philibert R, Dogan M, Beach SRH, Mills JA, Long JD (2020) AHRR methylation predicts smoking status and smoking intensity in both saliva and blood DNA. Am J Med Genet Part B Neuropsychiatr Genet 183(1):51–60. https://doi.org/10.1002/ajmg.b.32760

Poplin R et al (2018) A universal SNP and small-indel variant caller using deep neural networks. Nat Biotechnol 36(10):983–987. https://doi.org/10.1038/nbt.4235

Quang D, Xie X (2016) DanQ: a hybrid convolutional and recurrent deep neural network for quantifying the function of DNA sequences. Nucleic Acids Res 44(11):e107–e107. https://doi.org/10.1093/nar/gkw226

Ritchie MD et al (2013) Genome- and phenome-wide analyses of cardiac conduction identifies markers of arrhythmia risk. Circulation 127(13):1377–1385. https://doi.org/10.1161/CIRCULATIONAHA.112.000604

Sekhon A, Singh R, Qi Y (2018) DeepDiff: DEEP-learning for predicting DIFFerential gene expression from histone modifications. Bioinformatics 34(17):i891–i900. https://doi.org/10.1093/bioinformatics/bty612
Singh R, Lanchantin J, Robins G, Qi Y (2016) DeepChrome: deep-learning for predicting gene expression from histone modifications. Bioinformatics 32(17):i639–i648. https://doi.org/10.1093/bioinformatics/btw427
Smith AM et al (2020) Standard machine learning approaches outperform deep representation learning on phenotype prediction from transcriptomics data. BMC Bioinformatics 21(1):119. https://doi.org/10.1186/s12859-020-3427-8
Spinella J-F et al (2016) SNooPer: a machine learning-based method for somatic variant identification from low-pass next-generation sequencing. BMC Genomics 17(1):912. https://doi.org/10.1186/s12864-016-3281-2
Stetson LC, Pearl T, Chen Y, Barnholtz-Sloan JS (2014) Computational identification of multi-omic correlates of anticancer therapeutic response. BMC Genomics 15(7):S2. https://doi.org/10.1186/1471-2164-15-S7-S2
Tampuu A, Bzhalava Z, Dillner J, Vicente R (2019) ViraMiner: deep learning on raw DNA sequences for identifying viral genomes in human samples. PLoS One 14(9):e0222271. https://doi.org/10.1371/journal.pone.0222271
Thibodeau A, Uyar A, Khetan S, Stitzel ML, Ucar D (2018) A neural network based model effectively predicts enhancers from clinical ATAC-seq samples. Sci Rep 8(1):16048. https://doi.org/10.1038/s41598-018-34420-9
Thong Z, Tan JYY, Loo ES, Phua YW, Chan XLS, Syn CK-C (2021) Artificial neural network, predictor variables and sensitivity threshold for DNA methylation-based age prediction using blood samples. Sci Rep 11(1):1744. https://doi.org/10.1038/s41598-021-81556-2
Tian Q, Zou J, Tang J, Fang Y, Yu Z, Fan S (2019) MRCNN: a deep learning model for regression of genome-wide DNA methylation. BMC Genomics 20(2):192. https://doi.org/10.1186/s12864-019-5488-5
Trieu T, Martinez-Fundichely A, Khurana E (2020) DeepMILO: a deep learning approach to predict the impact of non-coding sequence variants on 3D chromatin structure. Genome Biol 21(1):79. https://doi.org/10.1186/s13059-020-01987-4
Turnbaugh PJ, Ley RE, Hamady M, Fraser-Liggett CM, Knight R, Gordon JI (2007) The human microbiome project. Nature 449(7164):804–810. https://doi.org/10.1038/nature06244
Vidaki A, Ballard D, Aliferi A, Miller TH, Barron LP, Syndercombe Court D (2017) DNA methylation-based forensic age prediction using artificial neural networks and next generation sequencing. Forensic Sci Int Genet 28:225–236. https://doi.org/10.1016/j.fsigen.2017.02.009
Wan Q, Pal R (2014) An Ensemble based top performing approach for NCI-DREAM drug sensitivity prediction challenge. PLoS One 9(6):e101183. https://doi.org/10.1371/journal.pone.0101183
Wang RY, Guo TQ, Li LG, Jiao JY, Wang LY (2020) Predictions of COVID-19 Infection Severity Based on Co-associations between the SNPs of Co-morbid Diseases and COVID-19 through Machine Learning of Genetic Data. In 2020 IEEE 8th International Conference on Computer Science and Network Technology (ICCSNT), pp 92–96
Wang Y, Franks JM, Whitfield ML, Cheng C (2019) BioMethyl: an R package for biological interpretation of DNA methylation data. Bioinformatics 35(19):3635–3641. https://doi.org/10.1093/bioinformatics/btz137
Wang Z, Gerstein M, Snyder M (2009) RNA-Seq: a revolutionary tool for transcriptomics. Nat Rev Genet 10(1):57–63. https://doi.org/10.1038/nrg2484
Wolf FA, Angerer P, Theis FJ (2018) SCANPY: large-scale single-cell gene expression data analysis. Genome Biol 19(1):15. https://doi.org/10.1186/s13059-017-1382-0
Wu H et al (2018) Metagenomics biomarkers selected for prediction of three different diseases in Chinese population. Biomed Res Int 2018:2936257. https://doi.org/10.1155/2018/2936257

Yachida S et al (2019) Metagenomic and metabolomic analyses reveal distinct stage-specific phenotypes of the gut microbiota in colorectal cancer. Nat Med 25(6):968–976. https://doi.org/10.1038/s41591-019-0458-7

Yang H, Wei Q, Li D, Wang Z (2020) Cancer classification based on chromatin accessibility profiles with deep adversarial learning model. PLoS Comput Biol 16(11):e1008405. https://doi.org/10.1371/journal.pcbi.1008405

Yin B et al (2019a) Using the structure of genome data in the design of deep neural networks for predicting amyotrophic lateral sclerosis from genotype. Bioinformatics 35(14):i538–i547. https://doi.org/10.1093/bioinformatics/btz369

Yin Q, Wu M, Liu Q, Lv H, Jiang R (2019b) DeepHistone: a deep learning approach to predicting histone modifications. BMC Genomics 20(2):193. https://doi.org/10.1186/s12864-019-5489-4

Yuan F, Lu L, Zou Q (2020) Analysis of gene expression profiles of lung cancer subtypes with machine learning algorithms. Biochim Biophys Acta - Mol Basis Dis 1866(8):165822. https://doi.org/10.1016/j.bbadis.2020.165822

Yuan V et al (2019) Accurate ethnicity prediction from placental DNA methylation data. Epigenetics Chromatin 12(1):51. https://doi.org/10.1186/s13072-019-0296-3

Zeng W, Wang Y, Jiang R (2019) Integrating distal and proximal information to predict gene expression via a densely connected convolutional neural network. Bioinformatics 36(2):496–503. https://doi.org/10.1093/bioinformatics/btz562

Zhang H et al (2019) VariFAST: a variant filter by automated scoring based on tagged-signatures. BMC Bioinformatics 20(22):713. https://doi.org/10.1186/s12859-019-3226-2

Zheng C, Xu R (2020) Predicting cancer origins with a DNA methylation-based deep neural network model. PLoS One 15(5):e0226461. https://doi.org/10.1371/journal.pone.0226461

Zhong H, Kim S, Zhi D, Cui X (2019) Predicting gene expression using DNA methylation in three human populations. PeerJ 7:e6757. https://doi.org/10.7717/peerj.6757

Zhou J, Troyanskaya OG (2015) Predicting effects of noncoding variants with deep learning–based sequence model. Nat Methods 12(10):931–934. https://doi.org/10.1038/nmeth.3547

Zhou Q, Jiang Q, Wei D (2011) A new method for classification in DNA sequence. In 2011 6th international conference on computer science education (ICCSE), pp 218–221

Zrimec J et al (2020) Deep learning suggests that gene expression is encoded in all parts of a co-evolving interacting gene regulatory structure. Nat Commun 11(1):6141. https://doi.org/10.1038/s41467-020-19921-4

How Machine Learning Has Revolutionized the Field of Cancer Informatics?

5

Ahmad Abu Turab Naqvi, Syed Afzal Murtaza Rizvi, and Md Imtaiyaz Hassan

Abstract

Cancer is among the most threatening maladies that humans are struggling to find a cure for ages. Greek physician and philosopher Hippocrates first introduced the world with the word cancer. Historical evidence suggests its presence tracing back to ~1500 BC. And we are still in the quest for finding a cure for this deadly disease despite all the medical and technological advances. If not curing completely, enhancing the chances of survival among cancer patients is also a dreaded goal for the scientific community. Early diagnosis and detection of the cancer type and its severity help in therapeutic intervention. Among all the technological inventions that have contributed to finding a cure for cancer, machine learning ever since its introduction to the medical sciences has been explored in various aspects of cancer research. From developing models for cancer detection, prognosis, and prediction of survival rate to the estimation of cancer recurrence, machine learning methods have been proved to provide hopefulness to the scientific community. In the chapter, we discuss machine learning applications in cancer research in various prospects like cancer detection/diagnosis, prognosis, susceptibility and risk assessment, prediction of survival rate, and predicting the chances of cancer recurrence. Besides providing the relevant case studies, we also discuss various machine learning techniques such as artificial neural networks (ANNs), support vector machine (SVM), decision tree and random forest (DT & RF), and Naïve Bayesian algorithms that are widely applied in cancer research. This chapter aims to give the reader a sense of how

A. A. T. Naqvi · S. A. M. Rizvi
Department of Computer Science, Jamia Millia Islamia, New Delhi, India

M. I. Hassan (✉)
Centre for Interdisciplinary Research in Basic Sciences, Jamia Millia Islamia, New Delhi, India
e-mail: mihassan@jmi.ac.in

S. Singh (ed.), *Machine Learning and Systems Biology in Genomics and Health*,
https://doi.org/10.1007/978-981-16-5993-5_5

machine learning has revolutionized cancer research by making mindful use of enormous cancer data being produced all over the globe.

5.1 Background

Cancer kills millions of people each year. Global statistics regarding cancer-related deaths are horrifying (Siegel et al. 2021). Despite so many advancements and diagnostic and therapeutic techniques, it's still among the most significant challenges in medical science. Various aspects of cancer-related approaches such as cancer diagnosis, prognosis, and treatment have improved significantly. However, disease heterogeneity and underlying molecular events leading to cancer, despite all the knowledge we have gathered so far by years-long efforts of biologists and medical scientists, are still not known up to that extent which would help develop effective management strategies for cancer.

Specialists working on cancer worldwide have resorted to different data sources for cancer diagnosis, which is undoubtedly the most challenging factor in cancer research and is considered essential for effective therapy and management. Developments in imaging and sequencing techniques have increased the production of enormous data for various cancer available through open-source databases and privatized data collections for researchers. Besides, a considerable amount of epidemiological and clinical data is also available for various cancers. Data sources such as epidemiological, genomic, mammographic, demographic, and clinical data have been vastly exploited for detecting and predicting various aspects of cancer like cancer susceptibility, recurrence, cancer survival, and identification of potential biomarkers of cancer.

Since its inception in the area of medical sciences, machine learning, a subfield of artificial intelligence, has been widely applied for diagnostic and prognostic purposes (Cruz and Wishart 2007; Kourou et al. 2015; Landhuis 2020) using a wide range of heterogeneous data sources described above. Cancer biology and cancer genomics are among the fields of science that have broadly benefitted from the powers of machine learning techniques for cancer diagnosis, prognosis, cancer classification, biomarker detection, prediction of recurrence, survival, and overall disease management. Machine learning algorithms can detect specific patterns in the data and make predictions based on those trends. Applications of machine learning algorithms in cancer research way back to almost three decades (Maclin et al. 1991; Simes 1985) and recent times have seen significant progress in the successful predictions using machine learning techniques. A major contributing factor to the increased applications of machine learning in cancer research is the availability of a considerable amount of cancer-related data since the success rate of machine learning-based predictions is proportional to the amount and health of data available.

This chapter discusses how machine learning has revolutionized cancer research and contributed to developing effective remedies for cancer. To make our point, we give a brief overview of machine learning and other technical aspects of it for the readers who are not or less familiar with the subject. We also discuss different data sources, genomic/clinical, used for machine learning-based predictions in cancer research. Finally, to further demonstrate the trends in machine learning applications in cancer research, we aim to present some representative and breakthrough studies as examples covering cancer research aspects.

5.2 Machine Learning Algorithms

As the name somehow reflects, machine learning gives the machine the ability to learn from data and make predictions. It is a subfield of artificial intelligence that mainly deals with making predictions in novel instances based on patterns/trends in the input data. Although various definitions are available for machine learning, let us stick to two definitions that help us understand the concept more formally. Renowned computer scientist and one of the pioneers of artificial intelligence Arthur Samuel, credited for popularizing the term machine learning in the 1960s, defines machine learning as "the field of study that gives computers the ability to learn without being explicitly programmed" (Samuel 2000). To learn without being explicitly programmed is the punch line that defines one of the distinctive features of machine learning. Learning from data and predicting trends in novel instances is the soul of machine learning. The definition of Arthur Samuel is self-explanatory and almost gives an idea of what machine learning is and how it differs from traditional computer programs to perform specific tasks. For a more technical and formal definition of machine learning, let us move to Tom Mitchell, an American computer scientist and an active researcher in the field of machine learning. By his definition of machine learning, "A computer program is said to learn from experience E concerning some class of tasks T and performance measure P, if its performance at tasks in T, as measured by P, improves with experience E" (Mitchell 1997). Unlike Arthur Samuel's straightforward definition, Mitchell's definition is considered more modern and formal.

Though reading through the definition in one go might sound like a tongue twister and less understandable, still, if we look at it through an example, we see that this definition is more explanatory than others. There are three terms that definition talks about: experience E, task T, and performance measure P. To make this clear, let us assume that we design a machine learning application to predict whether a tumor is benign or malignant based on its size using the images of tumors. In this case, task T is to differentiate between benign and malignant tumors. Feeding the algorithm with the images of both benign and malignant tumors is the experience E. After the training, our application should classify benign and malignant tumors when new images are given. The performance of our application will be measured by how well it can differentiate between both the tumors from new instances. The ratio of true and false predictions, in this case, will be the performance measure P. We can improve the performance of our application using the feedback from this performance measure by providing it with more precise and representative images of both tumors, eventually enhancing its experience so that its performance in task T improves.

Let us now discuss the types of machine learning methods and some leading algorithms applied broadly for solving various problems. However, while discussing the algorithms, we will restrict ourselves to those algorithms with practical applications in cancer research and relevant case studies.

Machine learning is divided into three major categories: (i) supervised learning, (ii) unsupervised learning, and (iii) reinforcement learning. Apart from these three, a fourth category named semi-supervised learning is also in practice based on the features of supervised and unsupervised methods where partial labeling or knowledge of desired outcomes is applied.

5.2.1 Supervised Learning

In this approach, the dataset fed to the algorithm contains preliminary information about the outcomes of the predictions, which is called the labeling of data. The algorithms aim to assign relative "class" or target output to the novel instances based on these labels. Besides, continuous numerical values can also be predicted using a known set of values. Classification and regression are two significant tasks performed using supervised machine learning methods. Neural networks, support vector machines, k-nearest neighbors, decision trees, random forests, and linear/logistic regression are known algorithms in this category.

5.2.2 Unsupervised Learning

In these methods, the data is not already labeled, unlike the supervised learning approach. Hence the algorithms aim to find a specific pattern in the data to group or categorize them. Clustering, anomaly detection, association rule learning, dimensionality reduction, etc. are typical unsupervised learning tasks.

5.2.3 Reinforcement Learning

Reinforcement learning is majorly associated with robotics, automated vehicles, and other such tasks. It is a typical environment-based approach where an intelligent agent designed to do a particular task, for example, housekeeping robots, acts to enhance reward or decrease penalty by learning and applying an optimized strategy.

Since we have described various types of machine learning methods and the kinds of tasks they are used to perform, let's dwell on the machine learning algorithms that belong to these categories. While looking for relevant literature on machine learning applications in cancer research, we found out that algorithms like artificial neural networks, support vector machines, naïve Bayes, decision trees, and random forests have been widely applied for cancer prediction, detection and prognosis using heterogeneous data. Therefore, before discussing case studies, let us have an idea of these machine learning algorithms.

5.3 Artificial Neural Networks (ANNs)

Neurons loosely inspire the idea of ANNs in the human brain, and the synaptic connections between neurons process information for proceeding signals to the connection of neurons for neurological functions. The origin of ANNs ways to back seminal works of McCulloch and Pits (McCulloch and Pitts 1943) and then later on the perceptron model by Rosenblatt (Rosenblatt 1961). ANNs are capable of solving problems where processing real-world continuous and discrete functions are in the subject. Interconnect units in ANNs are called neurons to take real-valued inputs and produce a single-valued output. This single-valued output might also be taken as input for another set of units as in multilayered ANNs. A simple ANN contains an input layer, a hidden layer, and an output layer that fires the generated output value. Based on the complexity of the input data and solving more complex problems, there might be ANNs with multiple hidden layers.

5.4 Support Vector Machines (SVMs)

SVMs (Cortes and Vapnik 1995) are machine learning models used for linear/nonlinear classification and regression tasks. SVMs are also capable of outlier detection in some instances. The two essential elements that majorly guide the functioning of SVMs are support vectors and decision boundaries. In the context of linearly separable instances for binary classification, SMVs work by constructing a hyperplane on the feature space to maximize the separation margin between both classes. The hyperplane formed in this way is called the decision boundary. Data points closest to the hyperplane, or decision boundary on which the optimal location of the hyperplane depends, are called support vectors.

5.5 Decision Trees and Random Forests

Decision trees (Breiman et al. 1984) are among the most potent learning methods for inductive learning purposes. They can deal with discrete-valued functions using the treelike structure to process input features and provide outputs. Due to their architectural simplicity, decision trees are easy and quick to learn and interpret. The attributes of input instances are represented through root nodes that offshoot branches with the values of given input instances where it further goes down. The possible output for that node is represented through the leaf node, and based on that, inference instances are classified, and predictions are made. Random forests are the ensembles of decision trees that further enhance their ability to make accurate predictions. While retaining the simplicity of decision trees, random forests can increase their predictive powers with various magnitudes.

5.6 Bayesian Networks

Having Bayes theorem as their foundation, Bayesian learning methods provide probabilistic solutions to machine learning problems. Naïve Bayes classifiers are the simplest classification models among the following Bayesian learning strategies. Bayesian networks (Pearl 1985) represent conditional probabilities of the instance variables using cyclic graphs.

5.7 Applications of Machine Learning in Cancer Research: Case Studies

So far, we have covered a brief introduction to machine learning in general and some machine learning methods widely applied in cancer research. This section discusses some case studies where machine learning methods have been applied for cancer detection, prognosis and survival analysis, etc. We searched literature databases and integrated platforms like PubMed, Scopus, and Web of Science for retrieving relevant literature. Search queries like "Machine Learning" AND "Cancer Detection" AND "Cancer Prediction" AND "Cancer Prognosis" etc. were used to search these databases. For maintaining the time relevance of the discussion, we only selected the papers published during the last 5 years. Most of these studies show machine learning methods for solving cancer research problems using various algorithms and a combination of heterogeneous data. Here, we present some representative case studies where the machine learning methods discussed in the previous section focus on problems like cancer prediction, prognosis, susceptibility, risk assessment, survival, cancer recurrence, and cancer classification. The terms prediction/prognosis are markers of identifying studies that mainly focus on susceptibility/risk assessment, survival prediction, and cancer recurrence prediction. Sometimes using various combinations of keywords helps to retrieve a wide range of literature covering the topic of concern compared to using specific keywords, which is less prone to skipping relevant literature. Therefore, we used the keywords prediction and prognosis, and the literature retrieved by this approach mainly focused on susceptibility, survival, and recurrence. Besides, cancer detection and classification are also significant aspects of this survey. Timely detection of cancer helps make effective predictive/prognostic strategies for raising the chances of remedy. Classification of cancer using expression or image data is also an aspect of cancer detection and helps identify the level and grade of cancer. Therefore, in this section, we present the case studies in cancer detection, classification, and then prediction/prognosis that will cover various aspects mentioned above.

5.8 Machine Learning in Cancer Detection

One of the significant factors that assure cancer therapy's effectiveness is the early detection of cancer in the patients. Unfortunately, in most cases, the cancer is diagnosed quite late, and the chances of treatment for such cancers are poor. In recent years, medical researchers have put significant efforts toward developing effective strategies for cancer detection, and applications of machine learning methods are one of them (Huang et al. 2018; Osareh and Shadgar 2010). Imaging, histopathological, and genomic data are primary sources for cancer detection using machine learning approaches. Let us dwell on some case studies where these data sources have been used with machine learning algorithms.

5.8.1 Breast Cancer Detection Using Mammographs

Detection of breast cancer using mammography is a standard approach. Identifying cancer using such images is a common practice among radiologists. However, in recent years, computer programs are being used for cancer detection using these images. While searching for machine learning applications in cancer detection, we came across two recent studies where machine learning models have been developed using the patient mammographs and algorithms trained to differentiate benign and malignant tumors and cancer staging. In a 2019 study, Shen Li and colleagues (Shen et al. 2019) used mammograms from the Digital Database for Screening Mammography (DDSM) and INbreast database to train convolutional neural networks learning-based algorithms for breast cancer detection. The datasets in this study were trained using CNNs following the "end-to-end" training approach, where lesion annotations were applied only for the initial stage of the training. The Area Under Curve (AUC) function was used to evaluate the trained models' prediction accuracy. The accuracy achieved by the model for both the datasets was found varying on AUC results. For the DDSA dataset, the single-model system achieved 0.88 (AUC) for each image whereas 0.91 (AUC) for the four-model averaged system. An AUC of 0.95 for a single model system was achieved for the INbreast dataset, and an average four model AUC was found 0.98. Another similar study was conducted by Kim et al. (Kim et al. 2020). The CNN model was trained using the mammographic data from five different institutes that comprised mammograms from cancer patients, for benign and normal samples. The key feature of this study is that it evaluated the algorithm's performance using the reader-based approach, where 14 radiologists from various centers volunteered to perform the detection with and without the assistance of a trained model. The performance was then evaluated in terms of AUC, sensitivity, and specificity. The overall performance of the standalone machine learning model was 0.95 (AUC). The model also performed well (AUC = 0.94) compared to the radiologists' prediction accuracy (AUC = 0.81). The authors also demonstrated that the radiologists' performance improved (AUC = 0.88) with the assistance of the trained model.

5.8.2 Breast Cancer Detection Using Histopathology Images

Other than mammograms, histopathology images are also significant for developing computational models for cancer detection. An ample amount of data for both cancer-positive and standard samples available in public repositories and institutional databases have facilitated the research community in exploring the possibilities of detecting cancer using machine learning techniques. However, in terms of applicability of a range of machine learning methods, most of these studies are limited to using CNNs for cancer detection and classification since CNN is more efficient in solving image processing problems. While searching for the relevant case studies for this particular section, we came across two recent studies which demonstrated the application of deep CNNs for breast cancer image classification. These studies have used the BreaKHis (Spanhol et al. 2016) dataset, a repository of ~8000 histopathological images collected from breast cancer patients containing benign and malignant tumor samples. The study conducted by Xie and colleagues (Xie et al. 2019) demonstrates the application of deep learning-based Inception models (Inception_V3 and Inception_ResNet_V2) (Szegedy et al. 2017) for image classification as well as clustering, which is an unsupervised approach. They also integrated the transfer learning technique with the models applied for the analysis. The analysis was performed on both original and augmented forms of the dataset. A binary and multiclass classification strategy was chosen for the classification of images. The results suggest that the Inception_ResNet_V2 model performs better than the other model as it can extract more expressive features from the image dataset. It is also observed that the overall classification accuracy of the augmented dataset was better in comparison to analyses done on the raw dataset because it is more viable to noise. The authors also demonstrated the limitation of the unsupervised clustering approach for image classification in comparison to the supervised method. In terms of the limitations of the overall analyses, noise and batch effect are found to be a significant deteriorating factor that affects the prediction accuracy and enhances the chances of producing false-positive outcomes. The other study conducted by Zhu et al. (Zhu et al. 2019) applied deep CNNs for breast cancer image classification. The authors used BreaKHis and the Breast Cancer Histology (BACH) (Aresta et al. 2019) datasets for the analysis. The key feature of this study is the application of "multiple compact CNNs" for image classification along with the hybrid CNN model approach. The results of the study suggest that the hybrid CNN model gives comparable classification accuracy for the BreaKHis dataset while the multi-model approach, for the BACH dataset, is claimed to outperform the state-of-the-art methods in overall prediction accuracy.

5.9 Machine Learning in Cancer Susceptibility and Risk Assessment

Cancer susceptibility prediction and risk assessment are crucial factors for strengthening the preventive and diagnostic measures. Machine learning has made significant improvements in the strategies to estimate the risk of developing cancer among individuals based on the clinical data and their exposure to various risk factors. In our quest for finding case studies for machine learning application in cancer susceptibility prediction, we came across various studies that use clinical data and various publicly available datasets for developing models that can efficiently predict the chances of a person developing cancer. Some of these studies have utilized available medical literature along with clinical data. A significant amount of these studies focused on breast cancer risk assessment using clinical and image data. For example, Ming et al. (2020) performed a risk assessment study on the clinical data taken from Geneva University Hospitals. They applied Markov Chain Monte Carlo generalized linear mixed model, adaptive boosting, and random forest classification.

Moreover, they compared the machine learning-based predictive model's performance with the previously available Breast and Ovarian Analysis of Disease Incidence and Carrier Estimation Algorithm (BOADICEA) model. AUC was used as the performance measure for prediction accuracy. The accuracy of machine learning-based models was higher (AUC = 0.88) compared to the BOADICEA model (AUC = 0.63).

Apart from patients' health records and other clinical data, a source like histopathological images, mammograms, and genomic profiles is also being explored to develop cancer risk prediction models using machine learning algorithms. During our survey, we came across an interesting study by the group of Yujia Bao (Bao et al. 2019) from the Massachusetts Institute of Technology, USA, and other experts from various institutes that developed cancer susceptibility prediction models using machine learning and natural language processing algorithms. Natural language processing is a subfield of artificial intelligence that mainly deals with developing models for speech recognition, text mining, robotics, etc. In this study, the authors conducted a literature survey on the PubMed database to retrieve relevant literature focusing on susceptibility genes and germline mutations for various cancers. Then, SVM and CNN algorithms were used to trained models for penetrance and prevalence classification using the annotated dataset retrieved from PubMed. For both tasks, the SVM-based model achieved an overall accuracy of 88.93% (penetrance) and 88.92% (prevalence). On the other hand, CNN-based models achieved 88.53% accuracy for penetrance classification whereas 88.52% accuracy for prevalence classification. The authors claim that the proposed strategy might help researchers in gene-cancer association studies, eventually affecting the decision-making process for clinical practices in cancer therapy.

5.10 Machine Learning in Cancer Recurrence Prediction

Recurrence, that is, coming back of cancer after treatment, is a frequent and challenging factor in cancer therapy. The rate and time of cancer recurrence can vary based on the type of cancer or site of origin. Therefore, dealing with cancer recurrence to enhance the effectiveness of the treatment has been a matter of concern for the oncologists. Several attempts have been made in recent times to explore the applications of machine learning techniques for predicting the recurrence of cancers. Experts have attempted to develop machine learning models using clinical, image, and genomic data for predicting cancer recurrence. In this section, we present some recent case studies where machine learning methods have been deployed for predicting the recurring tendencies of cancer in different pathological conditions. Xu et al. (2020) have presented an ensemble machine learning approach to predict stage IV colorectal cancer recurrence in patients who have undergone surgery for resection of the tumor site. To predict the chances of postoperative cancer recurrence, they applied logistic regression, decision tree, gradient boosting, and light-GBM algorithms on the dataset retrieved from the BioStudies database publicly available for researchers. The prediction accuracies of the trained models were measured based on AUC and F1 scores. The results suggest that among all four algorithms, gradient boosting performs well while acquiring an AUC value of 0.88 and F1 Score 0.912, which is the highest among all the algorithms applied. Lou and colleagues (2020), in a prospective cohort study aimed to predict the recurrence of breast cancer incidents within 10 years after the surgery. A dataset designed using the demographic attributes and clinical characteristics of the patient involved in the study were used to train, validate, and test the models. The authors also considered the quality of care given to the patients given during the 10 years' time as an attribute for prediction. Besides, the physical component summary score and the mental component summary score were also taken into account. This study is also interesting in terms of machine learning algorithms discussed in the previous section of the chapter. It applied algorithms like ANN, k-nearest neighbor (KNN), SVN, and naïve Bayesian model and Cox proportional-hazards regression model, a statistic model for prediction. Among all four models, as the results suggest, ANN scored highest in terms of specificity (95.52%), accuracy (94.12%), and AUC (0.97).

5.11 Conclusion

So far, we have discussed various aspects of machine learning applications in cancer research. The developments made during recent years in this domain give promising results and give glimpses of future advancements where machine learning is likely to reach levels that will enhance the chances of finding a cure for cancer. However, during our survey of relevant literature on the subject, we came across and realized certain limitations that leverage the effectiveness of these algorithms. As we know, machine learning algorithms work best when the data is healthy in terms of its accuracy and representativeness. Another challenge that we observed is the heterogeneity of the data sources and the lack of universal platforms that provide unified data from various sources specifically curated and organized for machine learning applications. The sensitivity of various machine learning algorithms toward data sparsity and heterogeneity is evident. However, working toward developing a universal data platform gathering all the cancer-specific data can be helpful and might pave the way for more efficient applications.

References

Aresta G, Araújo T, Kwok S, Chennamsetty SS, Safwan M, Alex V, Marami B, Prastawa M, Chan M, Donovan M, Fernandez G, Zeineh J, Kohl M, Walz C, Ludwig F, Braunewell S, Baust M, Vu QD, To MNN, Kim E, Kwak JT, Galal S, Sanchez-Freire V, Brancati N, Frucci M, Riccio D, Wang Y, Sun L, Ma K, Fang J, Kone I, Boulmane L, Campilho A, Eloy C, Polónia A, Aguiar P (2019) BACH: grand challenge on breast cancer histology images. Med Image Anal 56:122–139. https://doi.org/10.1016/j.media.2019.05.010

Bao Y, Deng Z, Wang Y, Kim H, Armengol VD, Acevedo F, Ouardaoui N, Wang C, Parmigiani G, Barzilay R, Braun D, Hughes KS (2019) Using machine learning and natural language processing to review and classify the medical literature on cancer susceptibility genes. JCO Clin Cancer Inform 3:1–9. https://doi.org/10.1200/CCI.19.00042

Breiman L, Friedman J, Olshen R, Stone C (1984) CART. Classification and regression trees. Wadsworth and Brooks/Cole, Monterey

Cortes C, Vapnik V (1995) Support-vector networks. Mach Learn 20(3):273–297. https://doi.org/10.1007/BF00994018

Cruz JA, Wishart DS (2007) Applications of machine learning in cancer prediction and prognosis. Cancer Inform 2:59–77

Huang S, Cai N, Pacheco PP, Narrandes S, Wang Y, Xu W (2018) Applications of support vector machine (SVM) learning in cancer genomics. Cancer Genomics Proteomics 15(1):41–51

Kim HE, Kim HH, Han BK, Kim KH, Han K, Nam H, Lee EH, Kim EK (2020) Changes in cancer detection and false-positive recall in mammography using artificial intelligence: a retrospective, multireader study. Lancet Digit Health 2(3):e138–e148. https://doi.org/10.1016/S2589-7500(20)30003-0

Kourou K, Exarchos TP, Exarchos KP, Karamouzis MV, Fotiadis DI (2015) Machine learning applications in cancer prognosis and prediction. Comput Struct Biotechnol J 13:8–17. https://doi.org/10.1016/j.csbj.2014.11.005

Landhuis E (2020) Deep learning takes on tumours. Nature 580(7804):551–553. https://doi.org/10.1038/d41586-020-01128-8

Lou SJ, Hou MF, Chang HT, Chiu CC, Lee HH, Yeh SJ, Shi HY (2020) Machine learning algorithms to predict recurrence within 10 years after breast cancer surgery: a prospective cohort study. Cancers (Basel) 12:12. https://doi.org/10.3390/cancers12123817

Maclin PS, Dempsey J, Brooks J, Rand J (1991) Using neural networks to diagnose cancer. J Med Syst 15(1):11–19

McCulloch WS, Pitts W (1943) A logical calculus of the ideas immanent in nervous activity. Bull Math Biophys 5(4):115–133. https://doi.org/10.1007/BF02478259

Ming C, Viassolo V, Probst-Hensch N, Dinov ID, Chappuis PO, Katapodi MC (2020) Machine learning-based lifetime breast cancer risk reclassification compared with the BOADICEA model: impact on screening recommendations. Br J Cancer 123(5):860–867. https://doi.org/10.1038/s41416-020-0937-0

Mitchell TM (1997) Machine learning. McGraw-Hill, New York

Osareh A, Shadgar B (2010) Machine learning techniques to diagnose breast cancer. 2010 5th international symposium on health informatics and bioinformatics

Pearl J (1985) Bayesian netwcrks: a model cf self-activated memory for evidential reasoning. Proceedings of the 7th conference of the cognitive science society, University of California, Irvine

Rosenblatt F (1961) Principles of neurodynamics. Perceptrons and the theory of brain mechanisms

Samuel AL (2000) Some studies in machine learning using the game of checkers. IBM J Res Develop 44:206–226. https://doi.org/10.1147/rd.441.0206

Shen L, Margolies LR, Rothstein JH, Fluder E, McBride R, Sieh W (2019) Deep learning to improve breast cancer detection on screening mammography. Sci Rep 9(1):12495. https://doi.org/10.1038/s41598-019-48995-4

Siegel RL, Miller KD, Fuchs HE, Jemal A (2021) Cancer statistics, 2021. CA Cancer J Clin 71(1):7–33. https://doi.org/10.3322/caac.21654

Simes RJ (1985) Treatment selection for cancer patients: application of statistical decision theory to the treatment of advanced ovarian cancer. J Chronic Dis 38(2):171–186

Spanhol FA, Oliveira LS, Petitjean C, Heutte L (2016) A dataset for breast cancer histopathological image classification. IEEE Trans Biomed Eng 63(7):1455–1462. https://doi.org/10.1109/TBME.2015.2496264

Szegedy C, Ioffe S, Vanhoucke V, Alemi A (2017) Inception-v4, inception-resnet and the impact of residual connections on learning. Proceedings of the AAAI Conference on Artificial Intelligence [Record #108 is using a reference type undefined in this output style]

Xie J, Liu R, Luttrell J t, Zhang C (2019) Deep learning based analysis of histopathological images of breast cancer. Front Genet 10:80. https://doi.org/10.3389/fgene.2019.00080

Xu Y, Ju L, Tong J, Zhou CM, Yang JJ (2020) Machine learning algorithms for predicting the recurrence of stage IV colorectal cancer after tumor resection. Sci Rep 10(1):2519. https://doi.org/10.1038/s41598-020-59115-y

Zhu C, Song F, Wang Y, Dong H, Guo Y, Liu J (2019) Breast cancer histopathology image classification through assembling multiple compact CNNs. BMC Med Inform Decis Mak 19(1):198. https://doi.org/10.1186/s12911-019-0913-x

Connecting the Dots: Using Machine Learning to Forge Gene Regulatory Networks from Large Biological Datasets. At the Intersection of GRNs: Where System Biology Meets Machine Learning

6

Isha Monga, Vinay Randhawa, and Sandeep Kumar Dhanda

Abstract

The last decade witnessed the exponential increase in the large and complex biological datasets. These studies range from human to different species under different conditions like disease versus control expression, evaluating clinical trials, microbiome, and metagenomics to generate a number of molecular entities like genes, proteins, transcription factors, and metabolites. The former ones still capture a gist of the rate at which computational repositories are getting filled up. These ever-increasing datasets provide an opportunity by raising a community question in the scientific field: about the effective utilization of the biological datasets for the predictive modeling to gain insights about the deep-seated meaningful information and hence to assist the health informatics community in tackling the grave diseases like autoimmune disorders, cancers, and childhood malignancies by predicting the better treatments or prior diagnosis. Therefore, since the last decade, the prime focus has been the understanding of the complex biological systems using the statistical and data analytical methods, which can specifically be made to handle large, heterogeneous, complex datasets in their training and yield some biological insight. Machine learning (ML) aims to address this question and provides predictive models which allow the scientific community to understand the disease from a new perspective and generate novel hypotheses. Present chapter is a primer on the use of machine learning to connect

I. Monga (✉)
Columbia University Irving Medical Center, New York, NY, USA

V. Randhawa
Department of Biochemistry, Panjab University, Chandigarh, India

S. K. Dhanda (✉)
Department of Oncology, St Jude Children's Research Hospital, Memphis, TN, USA
e-mail: Sandeep.Dhanda@stjude.org

S. Singh (ed.), *Machine Learning and Systems Biology in Genomics and Health*,
https://doi.org/10.1007/978-981-16-5993-5_6

the different dots hidden in the high-dimensional datasets and infer biological networks. It puts particular interest on the gene regulatory networks (GRNs) and the current computational resources available for the same. It provides a comprehensive picture of the GRNs along with the current advancements and limitations in the area.

6.1 Introduction

The central dogma of molecular biology is one of the greatest discoveries in biology linking the biochemical basis of a cell's transcription to one biomolecule to another (Crick 1970; Wang 2003). In a cell, DNA is transcribed to RNA, which in turn conveys the complex genetic code to the amino acids via transfer RNA (t-RNA) (Qu and Fang 2013; Brouwer and Lenstra 2019; Rodriguez and Larson 2020). Remarkable discovery of the noncoding genome along with the unexpected fact that the coding component only constitutes 1% of the whole mammalian genome discarded our old aged anthropocentric view toward proteins and considering noncoding genome as "junk DNA" (Ohno 1972; Collins 1997; Pennisi 2012; Thurman et al. 2012). The following decade was filled with the remarkable discoveries turning the so-called "junk DNA" into a treasure contributing various regulatory activities, hence it later termed as regulatory DNA (Bejerano et al. 2004; Ponting and Hardison 2011; Palazzo and Gregory 2014). The possibility of regulatory DNA opens up questions that gene expression is solely not dependent on the visible quantifiable gene products, but it reflects a complex system of interactions (Bartha et al. 2018; Brouwer and Lenstra 2019).

Thus, to fully picture the landscape of gene expression and regulation, one must not always focus on the quantitative gene expressions, but also how the quantitative gene blocks talk to one another in a complex temporal fashion (Ptashne 2014). Thus, the field of network biology came into picture by adopting the state-of-the-art statistical and mathematical approaches to understand and connect the dots of gene expression over the different time-points in the form of a network graph (Ocone et al. 2015).

However, it was evident that measuring interactions among genes is a more daunting task than measuring their abundances. The last two decades witnessed the growth in the gene-expression measurement technologies from microarrays to next-generation sequencing (NGS) advancements steadily decreasing costs; so much that the scientific community is able to infer the quantitative landscape of transcriptome in a temporal manner. Thus, availability of the time-point quantitative abundances provided a powerful insight to the scientific community to look for reconstructing or reengineering the iteration among genes computationally revealing the complex underpinning in gene expression patterns (Trapnell et al. 2014). Thus, in simple words, the scientific field which is dedicated to utilizing this connecting-the-dots approach to infer interactions collectively termed as network biology and networks is called gene regulatory networks (GRNs). In the broadest sense, a gene regulatory network (GRN) represents the causal relationship between genes that regulate the cellular/molecular functions (Mercatelli et al. 2020). The methods of GRN prediction attempt to recover the network topology based on the expression profiles of associated genes by using various statistical and computational methods (Huynh-Thu and Sanguinetti 2019; Mercatelli et al. 2020). In the present chapter, we covered the comprehensive roadmap till now in the area describing novel ways of identifying GRNs from multi-omics data for accelerating healthcare and personalized medicine.

This chapter is aimed to enhance the current understanding of biological researchers about recent advancements of machine learning-based methods for GRNs (see overall schematic workflow in Fig. 6.1). We start by leading the reader through a brief introduction on the current era, where vast amounts of biological data are being generated coupled with the need for better health informatics-based approaches to fuel precision and personalized medicine. We start off with detailed information on the current computational resources and databases in GRN inference in the first section. This is followed by a detailed second section covering a brief primer on machine learning, current methods (statistical and mathematical), and its role in the generation of regulatory networks by utilizing already available public biological data to reveal hidden meta-information. Here in this section, we discussed key terms, concepts, and current computational resources being utilized in regulatory network biology. Further, we included not only the gene based but extended our section by including the noncoding RNA centric GRNs.

Additionally, we provided some flavor to the reader about the current scenario of the deep learning-based tools and methods and their potential. We discussed how these next-generation advanced methods could be used to expand our understanding of complex biological networks and zoom into the disease biology canvas with a higher resolution. We finished the section by taking the reader into the practical illustrations where machine learning paved the way to decipher the novel targets from the complex networks for the therapeutic compounds in disease settings. In conclusion and future implications, we highlighted the current opportunities along with challenges on the way that need to be studied to fully realize the potential of machine and deep learning-based methods in GRN biology (Table 6.1).

6.2 Databases/Catalogues Covering GRN Information

In the start of this chapter, we would like to recognize the databases or catalogues that provide the curated information and help in building reliable machine learning models.

1. *TF2DNA*: TF2DNAis a very comprehensive resource for enlisting the binding specificities for transcription factor and their regulated genes (Pujato et al. 2014). In terms of coverage, TF2DNA focuses on the experimental studies on transcription factors for *Escherichia coli*, *Saccharomyces cerevisiae*, *Caenorhabditis*

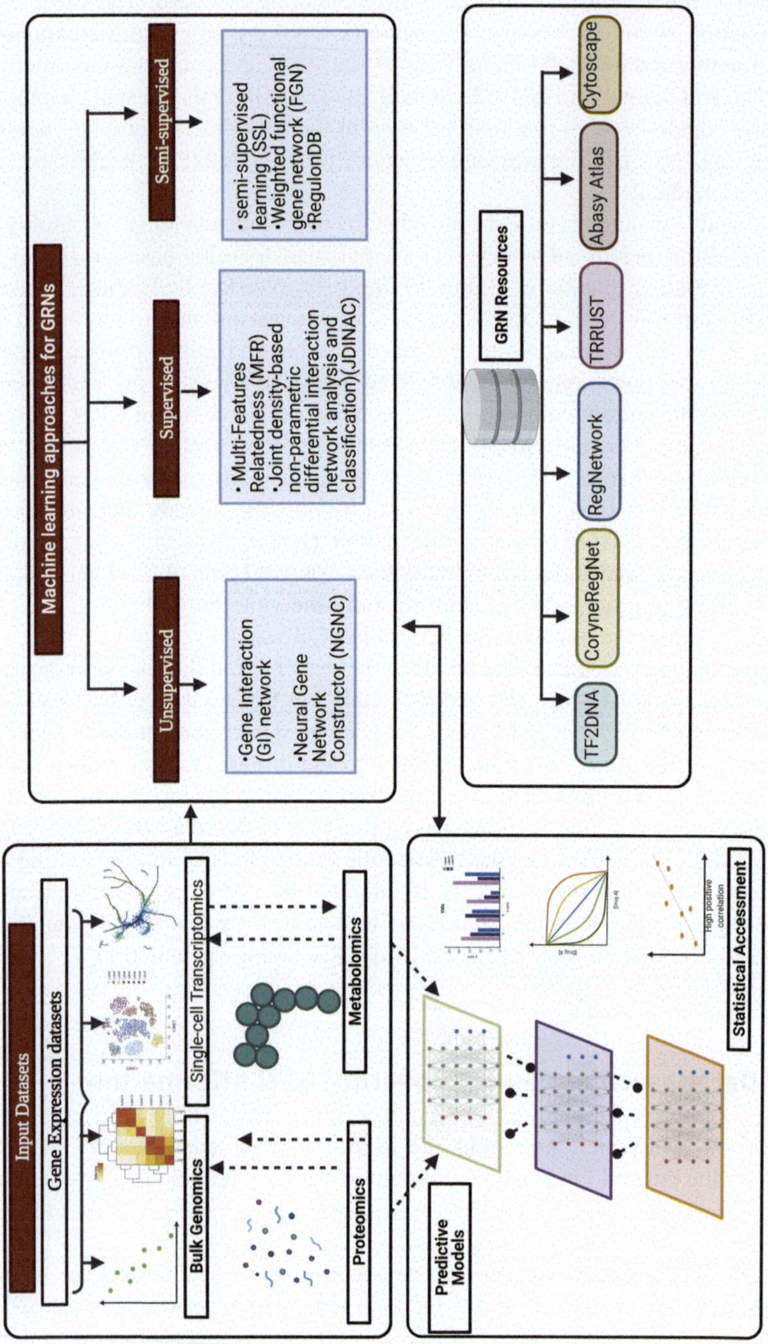

Fig. 6.1 Schematic illustration of the GRN input dataset types, methods, and resources covered in this chapter

Table 6.1 Software/tools available for the analysis and visualization of regulatory networks

Name	Summary	Website	Reference
Cytoscape	Cytoscape is an open-source software platform for visualizing complex networks and integrating these with any type of attribute data	http://cytoscape.org	Shannon et al. (2003), Cline et al. (2007)
Gephi	Gephi is an interactive visualization and exploration platform for all kinds of networks and complex systems, dynamic and hierarchical graphs	http://gephi.org/	Bastian et al. (2009)
Guess	GUESS is an exploratory data analysis and visualization tool for graphs and networks	http://graphexploration.cond.org	Pavlopoulos et al. (2017)
Graphviz	Graphviz is open-source graph visualization software. Graph visualization is a way of representing structural information as diagrams of abstract graphs and networks	http://www.graphviz.org/	Ellson et al. (2002)
Medusa	Medusa is a Java application for visualizing and manipulating graphs of interaction. It is available as an applet. It is an open source product under the GPL license	http://www.bork.embl.de/medusa	
BioLayout Express3D	BioLayout express 3D is released under the GNU public license (GPL) and is a tool for layout, visualization, and clustering of large scale networks	www.biolayout.org	
Osprey	Osprey, a stand-alone application for noncommercial use, is a software platform for visualization and manipulation of complex interaction networks	http://biodata.mshri.on.ca/osprey/servlet/Index	
ProViz	ProViz, a stand-alone open source application under the GPL license, is a tool for the visualization of protein- protein interaction networks	http://cbi.labri.fr/eng/provi z.htm	
Ondex	ONDEX, a stand-alone freely available open source application, is a database system that combines the features of semantic database integration and text mining with methods for graph-based analysis	http://www.ondex.org/	
PIVOT	PIVOT (protein interaction VisualizatiOn tool), a Java application, is a visualization tool for PPIs	http://www.cs.tau.ac.il/~rs hamir/pivot	

(continued)

Table 6.1 (continued)

Name	Summary	Website	Reference
Pajek	An open-source application provides analysis tools for large networks and graph-drawing capabilities, and it is free for noncommercial use. It runs under Windows OS only	http://vlado.fmf.uni-lj.si/pub/networks/pajek/	
NAViGaTOR	NAViGaTOR (network analysis, visualization, & graphing TORonto) is a software package for visualizing and analyzing protein-protein interaction networks	http://ophid.utoronto.ca/navigator/	Brown et al. (2009)
Cuttlefish	Cuttlefish runs on UNIX systems and employs many well-known layout algorithms for visualization. It supports interactive manipulation of layout and process visualization	http://cuttlefish.sourceforg e.net/	
SocNetV	Social network visualizer (SocNetV) is specifically a social network analysis tool that supports computation of statistics, centralities (node, closeness, betweenness, information, power), and a variety of layouts (leveled and radial)	http://socnetv.sourceforge. net/	
MeerKat	MeerKat is suitable for many types of network analysis, including social networks. If provides filtering mechanisms, interactive editing, support for dynamic networks and various metrics and automatically detects communities	http://www.aicml.ca/node/ 41	
Igraph and sna	R includes several packages relevant for social network analysis: Igraph: Generic network analysis package; sna: For sociometric analysis of networks; network manipulates and displays network objects	http://www.r-project.org/	
FlexGraph	FlexGraph is a distributed graph mining system can run very large graph computations on just a single machine	https://github.com/GraphChi	Park et al. (2020)
MMap	New software for fast scalable design and development of scalable algorithms	http://poloclub.gatech.edu/mmap/	Lin et al. (2014)

elegans, drosophila, mice, and humans. Additionally, the TF2DNA included theoretical studies for human transcription factors.

2. *CoryneRegNet*: The resource is aimed to compile the information on corynebacterial transcription factors and their gene regulatory networks (Pauling et al.

2012). In addition to providing the experimental data, CoryneRegNet also offers predicted networks (Baumbach 2010). The data is further explored in terms of clustering and re-annotations of selected networks (Baumbach et al. 2008; Wittkop et al. 2010).

3. *RegNetwork*: Developed in 2015, RegNetwork is an important repository to offer integrated information from transcriptional and posttranscriptional regulatory networks (Liu et al. 2015). It contains the regulatory relationship of transcription factors from mice and humans. The RegNetwork has sourced its content from many databases, like UniProt, KEGG, STRING, TransmiR, and mirRBase, and fetched data for transcription factors, microRNAs, and target genes.
4. *TRRUST*: TRRUST is another important resource for human and mouse transcriptional regulatory interactions. It was first published in 2015 (Han et al. 2015) and then updated in 2018 (Han et al. 2018). The database involves a manual curation for linking the transcription factors to their targets after initial text mining of research articles. In terms of number, the current version maintains the relationship of 1600+ transcription factors in humans and mice.
5. *Abasy Atlas*: Abasy Atlas is another comprehensive resource of regulatory networks in bacteria (Ibarra-Arellano et al. 2016; Escorcia-Rodriguez et al. 2020). The resource has listed 85 networks in nine bacterial species with 250,000+ regulatory interactions. In addition to providing meta-curated regulatory networks, Abasy Atlas offers predictions for global regulators, basal machinery genes, and modular and intermodular genes.
6. *AGRIS* (Arabidopsis Gene Regulatory Information Server): AGRIS includes information of *Arabidopsis* (*Arabidopsis thaliana*) transcription factors (TFs), predicted and experimentally verified *cis*-regulatory elements (CREs) and their interactions (Yilmaz et al. 2011). AGRIS is currently linked to three databases: At*cis*DB (*Arabidopsis thaliana* cis-regulatory database), AtTFDB (*Arabidopsis thaliana* transcription factor database), and AtRegNet (*Arabidopsis thaliana* regulatory network).
7. *Visualization Software*: Once the gene networks are either created or compiled from the literature or predicted using computational algorithms, the next challenge is to present these networks. Fortunately, we have a list of tools that could assist in visualizing these networks. In this section, we discussed some visualization apps to highlight connections in a regulatory network.
   - *Cytoscape*: Cytoscape is an open-source suite that can visualize not just regulatory gene networks but any short of complex networks with data attributes (Shannon et al. 2003). There is extensive support for apps built to fit in specific instances. At present, there are 11 apps with the "regulatory networks" tag. The most commonly used apps in these sections are:
     - ANIMO: analysis of networks with interactive modeling.
     - CyTargetLinker: to link/extend the biological networks or user networks with regulatory networks.
     - cyTransFinder: to reconstruct the signal transduction pathways from a group of genes,
     - iRegulon: to find out the master regulators and cis-regulatory interactions,

– Rene: to enhance regulatory information in a given KEGG/Reactome pathway.

NAViGaTOR: Network Analysis, Visualization, & Graphing TORonto (*NAViGaTOR*) is a collection of tools provided in the form of a user-friendly package for visualizing and analyzing networks. It supports 2D and 3D visualization of biological networks. The collection of tools includes algorithms for gene ontology (GO) and protein-protein interactions (PPIs) via inferring protein complexes visualized via different layouts like heatmap matrix, pheatmap, circular layouts, and protein directed graphs. NAViGaTOR is compatible with various community-driven data formats, namely, PSI-XML, BioPax, PAXtoolR, and GML (Brown et al. 2009).

6.3 Statistical Methods for GRN Construction

This section comprehends the state-of-the-art statistical methods along with a conceptual understanding behind these methods used to reverse engineer GRNs from time-series expression data. We discussed every features and concepts rather than a confusing reader with complex mathematical details. All the methods stated below are dependent on the time-series data, where the target variable can be mRNA or protein expression over various time-points for t1, t2, to tn. Similarly, the gene (mRNA) values over respective time derivatives can be encapsulated as (g1, g2 .... gn). Thus the value of each target gene/mRNA, that is, Yg, can then be written as a vector of Yg = $(yg,1,...,yg,T)^T$, where *Yg* = observation of target variable (g) at data point t.

Similarly, apart from the target gene variable observations at various time-points, there are other dependent factors, for instance, mRNA concentrations (say *x* here). These dependent factors can also be represented as the linear vector at different respective time-points ($x1^g$, $x2^g$,...xn^g). Further, there are mathematical procedures to infer the set of dependent regulators (summarized in detail here (Grzegorczyk et al. 2019).

1. *Gaussian Graphical Models (GGM)*: Penfold CA et al. developed a MATLAB-based software termed causal structure identification (CSI) utilizing the Gaussian process-based approach to predict GRNs from multiple time-series expression datasets. While CSI predicts a joint GRN from time-series data, within the algorithm, they utilize hierarchical approach to construct a separate GRN for each dataset thus favoring the identification of context-specific gene networks (Penfold et al. 2015). Similarly, Schäfer et al. proposed the graphical Gaussian models (GGMs) to infer conditionally dependent gene networks and hence gene association networks. Their method includes (1) better and regularized pool of small sample points of partial correlation; (2) they included a rigorous test, if they

include any edge in the GRN with an estimation of the degree of freedom for that particular node; and (3) they also included a system of heuristic network search to curb the FDR with multiple testing (Schafer and Strimmer 2005).

2. *Lasso and Elastic Net*: Gene networks in real biological datasets are very flexible, and earlier methods of capturing the gene expression like microarray and NGS are used to capture only few points of temporal gene expression of target gene (x). To keep this in mind, statisticians adopted the multivariate vector autoregressive (MVAR) methods, ridge, lasso, and elastic-net-based methods for GRN construction (van der Meijden et al. 2002). Due to inherent flexibility in these methods and recursive regularization (Shimamura et al. 2009), these methods are proven to be effective in stimulating GRNs even for small datasets and handling the FDR effectively. These methods have been used for GRN construction in the HeLa cell line (Whitfield et al. 2002).
3. *Hierarchical Bayesian Regression Models (HBR)*: In 2003, Husmeier et al. proposed the dynamic Bayesian networks for GRN inference utilizing the microarray gene expression data (Husmeier 2003). Since then, many variations of Bayesian networks are published in GRN biology field like Bayesian Orthogonal Least Squares (BOLS) (Kim 2007) and the following methods here (Grzegorczyk and Husmeier 2012; White and Vignes 2019).

While systems biologists work at the intersection of computational biology and statistics to develop statistical approaches, some of which are described above. However, these traditional statistical methods can be adopted to develop the computational tools and resources with user-friendly front ends or executable binaries/packages in GRN biology. Our chapter is dedicated on the computational tools developed utilizing these statistical approaches, and more detailed information on these methods is reviewed here (Penfold et al. 2015; Grzegorczyk et al. 2019).

6.4 Computational Tools and Methods for GRN Inference

The present section is aimed to catalogue some of the well-known computational tools for GRN construction. As already stated, a GRN represents the causal relationship between genes that regulate the cellular/molecular functions (Mercatelli et al. 2020). The methods of GRN prediction attempt to recover the network topology based on the expression profiles of associated genes by using various statistical and computational methods (Huynh-Thu and Sanguinetti 2019; Mercatelli et al. 2020). Thus, these statistical methods opted in the machine learning accelerated GRN inference form high-throughput biological data. Herein this section, we will shed light on machine learning and how its various categories have evolved in GRN biology over the last decade.

Machine learning (ML) techniques are generally implemented to address two major problems in GRNs: (1) obtaining the structure of a network from gene expression data and (2) assessing the characteristics and strength of interactions (Maetschke et al. 2014). These methods are generally implemented to address two major problems in GRNs: (1) obtaining the structure of the network from gene expression data and (2) assessing the characteristics and strength of interactions (Wiredja and Bebek 2017). We aimed to discuss the types of machine learning-based methods adopted for GRN inference under three main categories: *supervised, unsupervised, and semi-supervised algorithms*. Further, all these three types can also be exploited under time-series expression data. To make clear demarcation, we discussed the time-series GRN inference in a separate fourth section.

Various computational approaches have been developed to infer the regulatory networks from gene expression datasets. All these algorithms use a different set of rules to infer, validate, and implement the classification model. Based on the type of data available for network modeling, Patel et al. (Patel and Wang 2015) have categorized the existing gene regulatory network inference methods into three groups: (1) supervised, (2) unsupervised, and (3) semi-supervised (Patel and Wang 2015).

6.4.1 Supervised Algorithms

As term indicates, supervised learning is dependent on the training datasets and its associated features to map it on a certain set of outputs. Thus, based on the features of the training datasets, it predicts the outcome of an unknown target variable or dataset. The training datasets can be qualitative (with positive and negative training examples) or quantitative. The former comes under classification, while later belongs to regression-based training.

While positive examples could be easily obtained, negative examples are hard to find in most of the organisms. In general, an interaction between a pair of genes is considered to be known if there exists an experimental evidence of interaction, thus comprising the positive set. On the other hand, the invalidated gene-gene interactions comprise the negative dataset. Hence, if the output (Y) of the known input data (X) is already known, the algorithm learns the function to define their relationship as: Y = f(X). Multifeature relatedness (MFR) computes the relatedness between pair of genes by integrating similarity-based gene pair features (gene-expression level, GO annotation, homologous profile, and subcellular localization) and prior-knowledge-based (known gene-gene interactions from COXPRESdb, KEGG, and TRRUST databases) similarities. SVM (with the linear kernel) was implemented with associated learning algorithms for classification and regression analysis. MFR outperformed (AUC 0.819) in identifying gene-gene interactions compared to other linear models and co-expression analysis methods.

Joint density-based nonparametric differential interaction network analysis and classification (JDINAC), a neighborhood-based method for GRN prediction that computes the proximity between genes, on the basis of common neighbors in GI networks, was employed to identify the differential networks in the RNA-seq data having pathological diagnosis of AD. An optimal prediction model was developed (through cross-validation) that presented a good accuracy for AD prediction (Chen et al. 2019).

6.4.2 Unsupervised Algorithms

These algorithms infer networks solely based on the input data (X) and are independent of training examples, having no corresponding output variables. For example, GNE a deep learning framework was developed (to include biological information for the inference of gene interaction (GI) network) by integrating gene expression data and GI network topology information. Next, both topological properties and expression profiles were transformed into ID embedding (to preserve the topological structure proximity) and attribute embedding (to preserve the attribute proximity), respectively. Following, a multilayered ANN was implemented to ensemble the statistical relationships between topology and attribute information, for improving the GI predictions (Kc et al. 2019).

Neural gene network constructor (NGNC) is a deep learning model to infer gene regulatory networks from time-series gene expression data. It is a model-free heterogeneous model that is capable of reconstructing any network structure and could also predict the underlying dynamics. The methodology comprises two modules: network generators (by implementing Gumbel softmax technique) and a group of dynamics learners (consisting of several multilayer perceptron (MLPs), with each one corresponding to genes. The framework comprises (1) generation of the adjacency matrix, (2) computing element-wise products between gene expression vector X_t at time t and the column i of the adjacency matrix, (3) dynamic learning by considering the element-wise products (from (2)) and then computing the output X^i, which is an estimation of gene i's expression at time $t + 1$, (4) estimation of all gene expression values X_{t+1} at time $t + 1$ by an ensemble of outputs of all learners (Vohradsky 2001).

1. *Semi-Supervised Algorithms*: These algorithms are the intermediate between the supervised and unsupervised that utilize a positive-unlabeled learning approach. In supervised methods, some studies consider unknown interactions as negative examples that are not yet verified by experimental approaches; however, some of these unknown interactions might become positive examples in later experimental studies. For example, the number of regulatory interactions in RegulonDB (v 1.0) was increased from 533 to 4268 in RegulonDB (v 8.0), indicating validation of several functional interactions that were initially unknown in RegulonDB (Gama-Castro et al. 2016). Under this situation, a semi-supervised learning approach is generally more appropriate that considers all unknown interactions as unlabeled and utilizes positives and unlabeled for GRN inference (Patel and Wang 2015). In these methods, of the whole input data (X), only some data is labeled (Y), and algorithms usually begin with the labeled examples, followed by estimating the unlabeled data.

A graph-based semi-supervised learning (SSL) classifier, which is an intermediate between supervised and unsupervised learning, was developed to accurately predict pairwise synthetic genetic interactions (SGI) by integrating PPIs, protein complex, and gene expression data. The topological property of protein pairs in a weighted functional gene network (FGN) was considered as input features for the classification algorithm (You et al. 2010). SSL classifiers are comparatively more efficient than SVM, and this method could accurately identify (sensitivity 92%, specificity 91%) genetic interactions in yeast (You et al. 2010).

SVM and RF were employed in a semi-supervised manner to predict the GRN from gene expression profile data of *E. coli* and *S. cerevisiae*. During the analysis, an inductive (in which the model is learned upon the training data and is then applied to a separate disjoint testing dataset) and transductive (in which the model is learned upon training data and some information from the testing dataset and is then applied to the testing dataset) learning methods were used, both of which obtain negative training data from the unlabeled data iteratively. While evaluating the performance of these methods, the transductive learning approach outperformed the inductive learning approach in both the organisms. Importantly, experimental results also indicated semi-supervised methods to outperform the existing supervised methods (Patel and Wang 2015).

6.4.3 GRNs from the Time-Series Data

Measurements of gene expression are typically performed at a single time-point (termed steady-state) or during successive time-points (time-series), after some perturbations. While both data types have been used to infer GRNs, time-series data provides a more valuable information about network topology and dynamics of GRNs and is therefore considered to be more useful to reveal directional interactions among genes. Before 2016, it was considered that GRN inference methods from gene expression datasets are quite challenging; hence the DREAM (Dialogue on Reverse Engineering Assessment and Methods; http://dreamchallenges.org/) challenge (DREAM5-Network Inference Challenge [2010]) was offered to the open research community for reverse engineering the gene regulatory networks from gene expression datasets, utilizing "the wisdom of crowds."

Later, Marbach et al. (2012) compared 30 GRN inference methods (on *E. coli*, *Staphylococcus aureus*, *S. cerevisiae*, and in silico microarray data) submitted to the DREAM5 challenge. Authors observed that no single inference method performed optimally across all the data sets, while an integration of predictions from multiple inference methods provided a robust and high performance (Marbach et al. 2012). In another important study, Madhamshettiwar et al. (2012) also evaluated nine state-of-the-art gene regulatory network inference methods, encompassing the RFs, SVMs, MI, correlation, and partial correlation, on 38 simulated datasets (Madhamshettiwar et al. 2012), and then applied the best-performing method to predict normal and tumor networks. Authors observed large differences in the accuracy of developed methods that depend upon various factors, including data features, network size and topology, experiment type, and parameter settings. When the supervised method, Supervised Inference of Regulatory Networks (SIRENE) (Mordelet and Vert 2008), the best-performing method, was applied to a human serous papillary ovarian adenocarcinoma dataset, novel regulatory interactions were inferred.

However, during the last 5 years owing to the discovery of single-cell genomics, it's now possible for GRN construction with time-series data, and now GRN biology is well equipped to handle the time-series expression data not only form bulk transcriptomics (Micorarrays/RNAseq) but also from the resolution at a single-cell level. Below we described some of the tools generating the regulatory networks utilizing the single-cell gene expression datasets over various time-points:

1. *SCODE*: Present algorithm is designed to infer regulatory networks from single-cell RNA-Seq datasets. It utilizes Bayesian framework for the GRN inference from time and pseudo-time series data (Sanchez-Castillo et al. 2018).
2. *Monocle*: One of the most established and early tools to provide the concept of ordering the single-cell according to time-series gene expression snapshots termed as pseudo-time analysis (Trapnell et al. 2014). Apart from regulatory network guided form ordering the pseudo times, it offers users the opportunity to identify the rare or novel cell population and dissect the transcriptional heterogeneity into simpler states by simplifying ordering them according to pseudo time. Newer version is Monocle 2 implementing the pseudo time via reverse graph-embedding technique (Qiu et al. 2017).
3. *Slingshot*: A unique, easy to use tool for inferring cell lineages and pseudo time from single-cell data with the ability to identify multiple trajectories (Street et al. 2018).
4. *SCUBA*: Single-cell clustering using bifurcation analysis (SCUBA) is a robust, user-friendly, and efficient computational method to delineate the course of cell lineages from time-course scRNAseq data. It implements the concept of k-means clustering to cluster input datasets along a time-course gene expression pattern into a binary tree entailing bifurcation or cell lineage events (Marco et al. 2014).
5. Similarly, scNPF was designed for the preprocessing of scRNAseq data based on the concepts of network propagation and fusion (NPF). Multiple steps from gene expression to clustering and cell-type identification solely depend upon how the scRNAseq data is preprocessed. Its algorithms leverage the biological insightful knowledge with two computational steps: (a) making context-specific topology available in the given dataset and (b) combining it with the a priori information collected from publicly available molecular interaction networks to help understanding the network relationships in an efficient manner (Ye et al. 2019).

6.4.4 Feedforward Loops and Regulatory Networks

Among many genetic factors, transcription factors (TFs) and microRNAs (miRNAs) are the two most important regulators of gene expression, whose dysregulation may lead to various disease or distinct phenotypes (Monga and Kumar 2019). While TFs modulate gene transcription at the transcriptional level, miRNAs commonly regulate the gene expression at the posttranscriptional level (Hobert 2008; Monga and Banerjee 2019). For example, the development of stem cells is under the regulation of gene expression, miRNAs, and TFs (Gottgens 2015), where miRNAs are cell-intrinsic regulators of reprogramming and differentiation and TFs are regulators of stem cell aging (Poplineau et al. 2019). In fact, the underlying regulatory transcriptional network plays a crucial regulatory function in stem cell development (Wilkinson and Gottgens 2013; Gottgens 2015; Kosan and Godmann 2016).

TFs and miRNAs are also capable of regulating each other; when a common target gene is regulated by both of them, they define a network motif called feedforward loop (FFL). FFLs are important regulatory units that are building blocks of gene regulatory networks, which can be of 3/4/5 nodes (Bo et al. 2021). The 3-node FFL (consisting of TF, miRNA, and their mutual regulated gene) is one of the most common motif types in transcriptional network, which can be further extended to a 4-node FFL, where a co-expressed gene is included in 3-node FFL as a common target of TF and miRNA. Similarly, a 5-node FFL could also be established by including an miRNA-miRNA interaction to the 4-node FFL (Bo et al. 2021). All FFLs are classified into three major categories: (1) TF- FFL (where TF regulates miRNA and gene while miRNA represses target gene), (2) miRNA-FFL (where miRNA represses both TF and gene expression while TF regulates target gene expression), and (3) composite FFL (where TF regulates miRNA and target gene while miRNA represses both TF and gene). The clinical importance of FFLs could be attributed to the fact that dysregulated TF-miRNA-mediated FFLs have been characterized in several complex diseases including schizophrenia, myocardial infarction, and several cancers (Qin et al. 2020). While various studies have inferred transcriptional regulatory networks using the FFLs (Peng et al. 2013; Wang et al. 2017), the general strategy to first infer FFLs and then establish the underlying regulatory network is as follows:

1. *Identifying Genes, TFs and miRNAs, and Associated Regulatory Interactions*: The lists of candidate genes, TFs, and miRNAs for a particular state/condition can be obtained either from the literature, specific databases, or differential expression studies. The information of associated regulatory interactions is then obtained for regulatory databases (e.g., miRNA-gene/TF interactions, TF-miRNA, and TF-gene from miRTarBase (Hsu et al. 2011); VIRmiRNA (Qureshi et al. 2014); TransmiR (Tong et al. 2019); and TRRUST (Han et al. 2018), respectively).
2. *Identifying Statistically Significant Network Loops*: All interaction pairs are then unified and merged to infer FFLs, followed by assessing their statistical significance. For example, Lin et al. (2015) developed a statistical method to assess significance by permuting candidate TFs and miRNAs with random ones (Lin et al. 2015).
3. *Regulatory Network Construction and Analysis*: A comprehensive regulatory network is finally established by integrating all the identified statistically significant composite FFLs. Critically important candidate genes, TFs and miRNAs could be selected or prioritized on the basis of topological analysis of regulatory networks.

6.5 Conclusion and Limitations

The last two decades witnessed advancements in high-throughput technologies due to parallel innovation in the health informatics and mathematical and computational methods in network biology. The traditional reverse engineering from the time-series observations of the gene/protein expression data to infer a biological network is termed as gene regulatory networks (GRNs) inference, a critical step and one of the challenging problems in systems biology in systems biology. The present field is constantly evolving and equipping itself with novel methods in an effort to understand complexities of GRNs.

In the present chapter, we discussed a brief timeline of the resources used for GRNs along with the challenges posed to improve them in future. We discussed the revolutionary GRN applicability in the current high-throughput gene and protein technologies including but not limited to bulk/single-cell RNASeq, proteomics, and gene- to protein-based GRN inference via GSEA. However, owing to the constant evolution in the GRN systems biology methods, this chapter still presents a whistle-stop tour for GRN biology. We believe the growth in the current field is potentially based on the constant cross- hybridization and advancements in mathematics, statistics, and network biology. In conclusion, this chapter serves as a comprehensive primer on the methods for GRN biology in the current post-genomic era.

Acknowledgments Authors used Biorender institutional license to create the figure. IM and SKD conceived the project; IM and VR drafted the manuscript; and IM, VR, and SKD edited and reviewed the manuscript.

References

Bartha I, di Iulio J, Venter JC, Telenti A (2018) Human gene essentiality. Nat Rev Genet 19(1):51–62

Bastian M, Heymann S, Jacomy M (2009) Gephi: an open source software for exploring and manipulating networks

Baumbach J (2010) On the power and limits of evolutionary conservation--unraveling bacterial gene regulatory networks. Nucleic Acids Res 38(22):7877–7884

Baumbach J, Wittkop T, Weile J, Kohl T, Rahmann S (2008) MoRAine—a web server for fast computational transcription factor binding motif re-annotation. J Integr Bioinform 5:2

Bejerano G, Pheasant M, Makunin I, Stephen S, Kent WJ, Mattick JS, Haussler D (2004) Ultraconserved elements in the human genome. Science 304(5675):1321–1325

Bo C, Zhang H, Cao Y, Lu X, Zhang C, Li S, Kong X, Zhang X, Bai M, Tian K, Saitgareeva A, Lyaysan G, Wang J, Ning S, Wang L (2021) Construction of a TF-miRNA-gene feed-forward loop network predicts biomarkers and potential drugs for myasthenia gravis. Sci Rep 11(1):2416

Brouwer I, Lenstra TL (2019) Visualizing transcription: key to understanding gene expression dynamics. Curr Opin Chem Biol 51:122–129

Brown KR, Otasek D, Ali M, McGuffin MJ, Xie W, Devani B, Toch IL, Jurisica I (2009) NAViGaTOR: network analysis, visualization and graphing Toronto. Bioinformatics 25(24):3327–3329

Chen H, He Y, Ji J, Shi Y (2019) A machine learning method for identifying critical interactions between gene pairs in Alzheimer's disease prediction. Front Neurol 10:1162

Cline MS, Smoot M, Cerami E, Kuchinsky A, Landys N, Workman C, Christmas R, Avila-Campilo I, Creech M, Gross B, Hanspers K, Isserlin R, Kelley R, Killcoyne S, Lotia S, Maere S, Morris J, Ono K, Pavlovic V, Pico AR, Vailaya A, Wang PL, Adler A, Conklin BR, Hood L, Kuiper M, Sander C, Schmulevich I, Schwikowski B, Warner GJ, Ideker T, Bader GD (2007) Integration of biological networks and gene expression data using Cytoscape. Nat Protoc 2(10):2366–2382

Collins FS (1997) Sequencing the human genome. Hosp Pract 32(1):53–34

Crick F (1970) Central dogma of molecular biology. Nature 227(5258):561–563

Ellson J, Gansner E, Koutsofios L, North SC, Woodhull G (2002) Graphviz - open source graph drawing tools. Graph Draw 2265:483–484

Escorcia-Rodriguez JM, Tauch A, Freyre-Gonzalez JA (2020) Abasy atlas v2.2: the most comprehensive and up-to-date inventory of meta-curated, historical, bacterial regulatory networks, their completeness and system-level characterization. Comput Struct Biotechnol J 18:1228–1237

Gama-Castro S, Salgado H, Santos-Zavaleta A, Ledezma-Tejeida D, Muniz-Rascado L, Garcia-Sotelo JS, Alquicira-Hernandez K, Martinez-Flores I, Pannier L, Castro-Mondragon JA, Medina-Rivera A, Solano-Lira H, Bonavides-Martinez C, Perez-Rueda E, Alquicira-Hernandez S, Porron-Sotelo L, Lopez-Fuentes A, Hernandez-Koutoucheva A, Del Moral-Chavez V, Rinaldi F, Collado-Vides J (2016) RegulonDB version 9.0: high-level integration of gene regulation, coexpression, motif clustering and beyond. Nucleic Acids Res 44(D1):D133–D143

Gottgens B (2015) Regulatory network control of blood stem cells. Blood 125(17):2614–2620

Grzegorczyk M, Aderhold A, Husmeier D (2019) Overview and evaluation of recent methods for statistical inference of gene regulatory networks from time series data. Methods Mol Biol 1883:49–94

Grzegorczyk M, Husmeier D (2012) A non-homogeneous dynamic Bayesian network with sequentially coupled interaction parameters for applications in systems and synthetic biology. Stat Appl Genet Mol Biol 11:4

Han H, Cho JW, Lee S, Yun A, Kim H, Bae D, Yang S, Kim CY, Lee M, Kim E, Lee S, Kang B, Jeong D, Kim Y, Jeon HN, Jung H, Nam S, Chung M, Kim JH, Lee I (2018) TRRUST v2: an expanded reference database of human and mouse transcriptional regulatory interactions. Nucleic Acids Res 46(D1):D380–D386

Han H, Shim H, Shin D, Shim JE, Ko Y, Shin J, Kim H, Cho A, Kim E, Lee T, Kim H, Kim K, Yang S, Bae D, Yun A, Kim S, Kim CY, Cho HJ, Kang B, Shin S, Lee I (2015) TRRUST: a reference database of human transcriptional regulatory interactions. Sci Rep 5:11432

Hobert O (2008) Gene regulation by transcription factors and microRNAs. Science 319(5871):1785–1786

Hsu SD, Lin FM, Wu WY, Liang C, Huang WC, Chan WL, Tsai WT, Chen GZ, Lee CJ, Chiu CM, Chien CH, Wu MC, Huang CY, Tsou AP, Huang HD (2011) miRTarBase: a database curates experimentally validated microRNA-target interactions. Nucleic Acids Res 39:D163–D169

Husmeier D (2003) Sensitivity and specificity of inferring genetic regulatory interactions from microarray experiments with dynamic Bayesian networks. Bioinformatics 19(17):2271–2282

Huynh-Thu VA, Sanguinetti G (2019) Gene regulatory network inference: an introductory survey. Methods Mol Biol 1883:1–23

Ibarra-Arellano MA, Campos-Gonzalez AI, Trevino-Quintanilla LG, Tauch A, Freyre-Gonzalez JA (2016) Abasy atlas: a comprehensive inventory of systems, global network properties and systems-level elements across bacteria. Database (Oxford) 2016:5

Kc K, Li R, Cui F, Yu Q, Haake AR (2019) GNE: a deep learning framework for gene network inference by aggregating biological information. BMC Syst Biol 13:38

Kim CS (2007) Bayesian orthogonal least squares (BOLS) algorithm for reverse engineering of gene regulatory networks. BMC Bioinformatics 8:251

Kosan C, Godmann M (2016) Genetic and epigenetic mechanisms that maintain hematopoietic stem cell function. Stem Cells Int 2016:5178965

Lin Y, Zhang Q, Zhang HM, Liu W, Liu CJ, Li Q, Guo AY (2015) Transcription factor and miRNA co-regulatory network reveals shared and specific regulators in the development of B cell and T cell. Sci Rep 5:15215

Lin Z, Kahng M, Sabrin KM, Chau DH, Lee H, Kang U (2014) MMap: fast billion-scale graph computation on a PC via memory mapping. Proc IEEE Int Conf Big Data 2014:159–164

Liu ZP, Wu C, Miao H, Wu H (2015) RegNetwork: an integrated database of transcriptional and post-transcriptional regulatory networks in human and mouse. Database (Oxford) 2015:95

Madhamshettiwar PB, Maetschke SR, Davis MJ, Reverter A, Ragan MA (2012) Gene regulatory network inference: evaluation and application to ovarian cancer allows the prioritization of drug targets. Genome Med 4(5):41

Maetschke SR, Madhamshettiwar PB, Davis MJ, Ragan MA (2014) Supervised, semi- supervised and unsupervised inference of gene regulatory networks. Brief Bioinform 15(2):195–211

Marbach D, Costello JC, Kuffner R, Vega NM, Prill RJ, Camacho DM, Allison KR, Consortium D, Kellis M, Collins JJ, Stolovitzky G (2012) Wisdom of crowds for robust gene network inference. Nat Methods 9(8):796–804

Marco E, Karp RL, Guo G, Robson P, Hart AH, Trippa L, Yuan GC (2014) Bifurcation analysis of single-cell gene expression data reveals epigenetic landscape. Proc Natl Acad Sci U S A 111(52):E5643–E5650

Mercatelli D, Scalambra L, Triboli L, Ray F, Giorgi FM (2020) Gene regulatory network inference resources: a practical overview. Biochim Biophys Acta Gene Regul Mech 1863(6):194430

Monga I, Banerjee I (2019) Computational identification of piRNAs using features based on RNA sequence, structure, thermodynamic and physicochemical properties. Curr Genomics 20(7):508–518

Monga I, Kumar M (2019) Computational resources for prediction and analysis of functional miRNA and their Targetome. Methods Mol Biol 1912:215–250

Mordelet F, Vert JP (2008) SIRENE: supervised inference of regulatory networks. Bioinformatics 24(16):i76–i82

Ocone A, Haghverdi L, Mueller NS, Theis FJ (2015) Reconstructing gene regulatory dynamics from high-dimensional single-cell snapshot data. Bioinformatics 31(12):i89–i96

Ohno S (1972) So much "junk" DNA in our genome. Brookhaven Symp Biol 23:366–370

Palazzo AF, Gregory TR (2014) The case for junk DNA. PLoS Genet 10(5):e1004351

Park C, Park HM, Kang U (2020) FlexGraph: flexible partitioning and storage for scalable graph mining. PLoS One 15(1):e0227032

Patel N, Wang JT (2015) Semi-supervised prediction of gene regulatory networks using machine learning algorithms. J Biosci 40(4):731–740

Pauling J, Rottger R, Tauch A, Azevedo V, Baumbach J (2012) CoryneRegNet 6.0--updated database content, new analysis methods and novel features focusing on community demands. Nucleic Acids Res 40:D610–D614

Pavlopoulos GA, Paez-Espino D, Kyrpides NC, Iliopoulos I (2017) Empirical comparison of visualization tools for larger-scale network analysis. Adv Bioinfo 2017:1278932

Penfold CA, Shifaz A, Brown PE, Nicholson A, Wild DL (2015) CSI: a nonparametric Bayesian approach to network inference from multiple perturbed time series gene expression data. Stat Appl Genet Mol Biol 14(3):307–310

Peng C, Wang M, Shen Y, Feng H, Li A (2013) Reconstruction and analysis of transcription factor-miRNA co-regulatory feed-forward loops in human cancers using filter-wrapper feature selection. PLoS One 8(10):e78197

Pennisi E (2012) Genomics. ENCODE project writes eulogy for junk DNA. Science 337(6099):1159–1161

Ponting CP, Hardison RC (2011) What fraction of the human genome is functional? Genome Res 21(11):1769–1776

Poplineau M, Vernerey J, Platet N, N'Guyen L, Herault L, Esposito M, Saurin AJ, Guilouf C, Iwama A, Duprez E (2019) PLZF limits enhancer activity during hematopoietic progenitor aging. Nucleic Acids Res 47(9):4509–4520

Ptashne M (2014) The chemistry of regulation of genes and other things. J Biol Chem 289(9):5417–5435

Pujato M, Kieken F, Skiles AA, Tapinos N, Fiser A (2014) Prediction of DNA binding motifs from 3D models of transcription factors; identifying TLX3 regulated genes. Nucleic Acids Res 42(22):13500–13512

Qin G, Mallik S, Mitra R, Li A, Jia P, Eischen CM, Zhao Z (2020) MicroRNA and transcription factor co-regulatory networks and subtype classification of seminoma and non-seminoma in testicular germ cell tumors. Sci Rep 10(1):852

Qiu X, Mao Q, Tang Y, Wang L, Chawla R, Pliner HA, Trapnell C (2017) Reversed graph embedding resolves complex single-cell trajectories. Nat Methods 14(10):979–982

Qu H, Fang X (2013) A brief review on the human encyclopedia of DNA elements (ENCODE) project. Genomics Proteomics Bioinformatics 11(3):135–141

Qureshi A, Thakur N, Monga I, Thakur A, Kumar M (2014) VIRmiRNA: a comprehensive resource for experimentally validated viral miRNAs and their targets. Database (Oxford) 2014:20140101

Rodriguez J, Larson DR (2020) Transcription in living cells: molecular mechanisms of bursting. Annu Rev Biochem 89:189–212

Sanchez-Castillo M, Blanco D, Tienda-Luna IM, Carrion MC, Huang Y (2018) A Bayesian framework for the inference of gene regulatory networks from time and pseudo-time series data. Bioinformatics 34(6):964–970

Schafer J, Strimmer K (2005) An empirical Bayes approach to inferring large-scale gene association networks. Bioinformatics 21(6):754–764

Shannon P, Markiel A, Ozier O, Baliga NS, Wang JT, Ramage D, Amin N, Schwikowski B, Ideker T (2003) Cytoscape: a software environment for integrated models of biomolecular interaction networks. Genome Res 13(11):2498–2504

Shimamura T, Imoto S, Yamaguchi R, Fujita A, Nagasaki M, Miyano S (2009) Recursive regularization for inferring gene networks from time-course gene expression profiles. BMC Syst Biol 3:41

Street K, Risso D, Fletcher RB, Das D, Ngai J, Yosef N, Purdom E, Dudoit S (2018) Slingshot: cell lineage and pseudotime inference for single-cell transcriptomics. BMC Genomics 19(1):477

Thurman RE, Rynes E, Humbert R, Vierstra J, Maurano MT, Haugen E, Sheffield NC, Stergachis AB, Wang H, Vernot B, Garg K, John S, Sandstrom R, Bates D, Boatman L, Canfield TK, Diegel M, Dunn D, Ebersol AK, Frum T, Giste E, Johnson AK, Johnson EM, Kutyavin T, Lajoie B, Lee BK, Lee K, London D, Lotakis D, Neph S, Neri F, Nguyen ED, Qu H, Reynolds AP, Roach V, Safi A, Sanchez ME, Sanyal A, Shafer A, Simon JM, Song L, Vong S, Weaver M, Yan Y, Zhang Z, Zhang Z, Lenhard B, Tewari M, Dorschner MO, Hansen RS, Navas PA, Stamatoyannopoulos G, Iyer VR, Lieb JD, Sunyaev SR, Akey JM, Sabo PJ, Kaul R, Furey TS, Dekker J, Crawford GE, Stamatoyannopoulos JA (2012) The accessible chromatin landscape of the human genome. Nature 489(7414):75–82

Tong Z, Cui Q, Wang J, Zhou Y (2019) TransmiR v2.0: an updated transcription factor- microRNA regulation database. Nucleic Acids Res 47(D1):D253–D258

Trapnell C, Cacchiarelli D, Grimsby J, Pokharel P, Li S, Morse M, Lennon NJ, Livak KJ, Mikkelsen TS, Rinn JL (2014) The dynamics and regulators of cell fate decisions are revealed by pseudotemporal ordering of single cells. Nat Biotechnol 32(4):381–386

van der Meijden CM, Lapointe DS, Luong MX, Peric-Hupkes D, Cho B, Stein JL, van Wijnen AJ, Stein GS (2002) Gene profiling of cell cycle progression through S-phase reveals sequential expression of genes required for DNA replication and nucleosome assembly. Cancer Res 62(11):3233–3243

Vohradsky J (2001) Neural network model of gene expression. FASEB J 15(3):846–854

Wang H, Luo J, Liu C, Niu H, Wang J, Liu Q, Zhao Z, Xu H, Ding Y, Sun J, Zhang Q (2017) Investigating MicroRNA and transcription factor co-regulatory networks in colorectal cancer. BMC Bioinformatics 18(1):388

Wang ZZ (2003) Central dogma of molecular biology. Sheng Li Ke Xue Jin Zhan 34(2):101–103

White A, Vignes M (2019) Causal queries from observational data in biological systems via Bayesian networks: an empirical study in small networks. Methods Mol Biol 1883:111–142

Whitfield ML, Sherlock G, Saldanha AJ, Murray JI, Ball CA, Alexander KE, Matese JC, Perou CM, Hurt MM, Brown PO, Botstein D (2002) Identification of genes periodically expressed in the human cell cycle and their expression in tumors. Mol Biol Cell 13(6):1977–2000

Wilkinson AC, Gottgens B (2013) Transcriptional regulation of haematopoietic stem cells. Adv Exp Med Biol 786:187–212

Wiredja D, Bebek G (2017) Identifying gene interaction networks. Methods Mol Biol 1666:539–556

Wittkop T, Emig D, Lange S, Rahmann S, Albrecht M, Morris JH, Bocker S, Stoye J, Baumbach J (2010) Partitioning biological data with transitivity clustering. Nat Methods 7(6):419–420

Ye W, Ji G, Ye P, Long Y, Xiao X, Li S, Su Y, Wu X (2019) scNPF: an integrative framework assisted by network propagation and network fusion for preprocessing of single-cell RNA-seq data. BMC Genomics 20(1):347

Yilmaz A, Mejia-Guerra MK, Kurz K, Liang X, Welch L, Grotewold E (2011) AGRIS: the Arabidopsis gene regulatory information server, an update. Nucleic Acids Res 39:D1118–D1122

You ZH, Yin Z, Han K, Huang DS, Zhou X (2010) A semi-supervised learning approach to predict synthetic genetic interactions by combining functional and topological properties of functional gene network. BMC Bioinformatics 11:343

Identification of Novel Noncoding RNAs in Plants by Big Data Analysis

7

Mamun Mandal, Nikita Poddar, and Shailesh Kumar

Abstract

Noncoding RNAs (ncRNAs) include different types of molecules which are not involved in the coding of proteins. Globally, different types of research are going on with ncRNAs in different types of organisms. In plants, ncRNAs have been deemed as crucial regulators of diverse types of biological processes such as development, vernalization, growth, homeostasis, and biotic and abiotic stress conditions. Current progress in experimentation and computational technologies has created fabulous momentum for the discovery of novel ncRNAs and their functional characterization. These new scientific establishments can be complimented by generating new databases and also update the present databases of ncRNAs. For this purpose, big data analysis and machine learning algorithms play a significant role. Nowadays, ncRNAs are an essential class of RNA having few appropriate databases and prediction tools. Here, we are providing the comprehensive information about the different classes of ncRNAs, their functional characters, and the applications. Additionally, we have summarized the information of different types of identification tools and databases for plant ncRNAs.

M. Mandal
Microbiology and Microbial Biotechnology Laboratory, Department of Botany, University of Gour Banga, Malda, West Bengal, India

N. Poddar · S. Kumar (✉)
Bioinformatics Lab, National Institute of Plant Genome Research (NIPGR), New Delhi, India
e-mail: shailesh@nipgr.ac.in

S. Singh (ed.), *Machine Learning and Systems Biology in Genomics and Health*,
https://doi.org/10.1007/978-981-16-5993-5_7

7.1 Introduction

RNA is the coding template for producing the proteins, essential for cellular functions. From the high perception level, RNA is considered to be either noncoding or coding. An unexpected invention from the genomic sequence annotations is that protein-coding sequences occupy a very small space of the genomic region (Liu et al. 2015). Most of the protein-coding RNA functionality is known and has been extensively studied due to its crucial function in biological regulation. Conversely, the functionality of the noncoding RNA (ncRNA) remains an ambiguity. They have been shown to play a great role in the regulation of development, growth, and stress condition of the plant at both transcriptional and posttranscriptional levels (Matsui et al. 2013; Sunkar et al. 2012). The ncRNAs show expression depending upon factors, such as cell function, type, and phenomenon. Previously the ncRNAs were known as "junk DNA" that contributes in many significant roles toward epigenomic regulation, gene transcription, and protein-coding gene expression (Wei et al. 2017). Based on molecular structure, plant ncRNAs can be divided into two classes: one is linear ncRNAs, and another one is circular ncRNAs (circRNAs; Liu et al. 2017). On the basis of molecular functional character, linear ncRNAs can be divided as housekeeping ncRNAs (such as small nucleolar RNAs, transfer RNAs, and ribosomal RNAs) and regulatory ncRNAs. However, on the basis of length, regulatory ncRNAs have been classified into two subcategories: (1) small noncoding RNAs (sncRNA) and (2) long noncoding RNAs (lncRNA), including sense transcripts, intergenic lncRNAs, and intronic lncRNAs, naturally occurring antisense transcripts (lnc-NATs); they regulate the gene expression level via different molecular mechanisms (Matsui et al. 2013; Liu et al. 2012; Santosh et al. 2014). The sncRNA can be alienated into many classes such as microRNAs (miRNAs), promoter-associated RNAs (PARs), small nuclear RNAs (snRNAs), small rDNA-derived RNA (srRNA), enhancer RNAs (eRNAs), tRNA-derived small RNA (tsRNA), PIWI-interacting RNAs (piRNAs), and small interfering RNA (siRNA; Axtell et al. 2013). Based on the genomic region from where circRNAs evolve, circRNAs can be divided as (a) intronic circRNAs, (b) exoniccircRNAs, (c) intergenic circRNAs, (d) UTR circRNAs, and (e) other circRNAs arising from two or more genes (Ye et al. 2015). The taxonomy of the ncRNA classes and their subclasses is shown in Fig. 7.1.

Although identification, functional characterization, and discovery of ncRNAs have been revealed half a centenary before, their broad phenomena and varied function in several plants have not been truthfully established till the post-genomics era (Holley et al. 1965). The advancement of next-generation sequencing (NGS) technology has led to an exploration of the high-throughput ncRNA seq data; consequently, it is becoming easier for us to find novel ncRNAs (Pauli et al. 2015; Li et al. 2014). One of the biggest challenges is the detection of novel ncRNAs in the coding potential determination of RNA sequences. The advent of ribosome sequencing (also known as ribosome profiling) has shown efficacy for discriminating between transcripts data of coding and noncoding sequences. Consequently, it ameliorates the accuracy of the ncRNA annotation data (Bazzini et al. 2014; Pauli

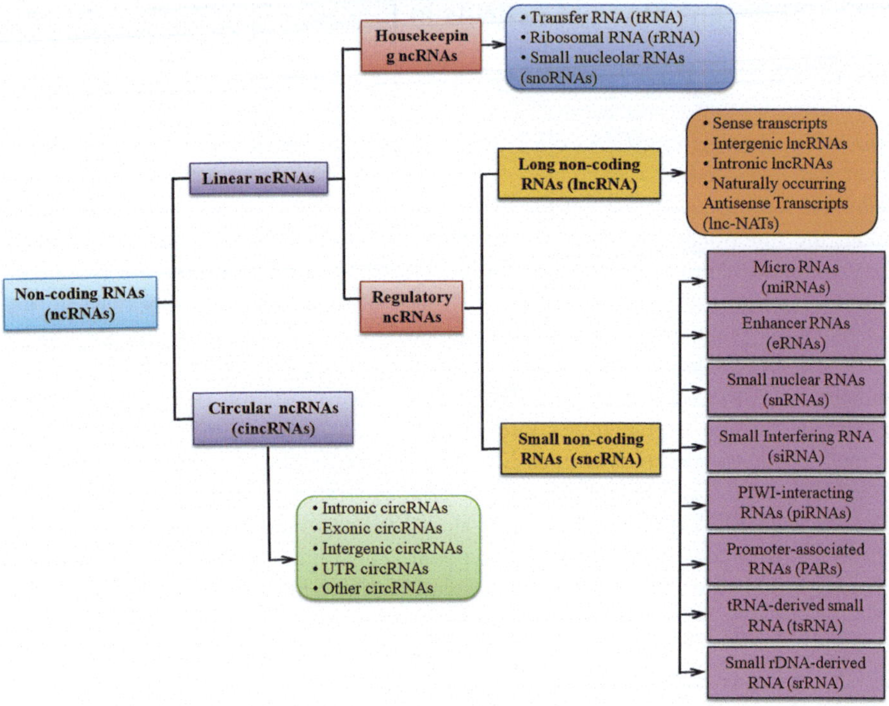

Fig. 7.1 Different types of noncoding RNAs

et al. 2014). Molecular genome engineering is a powerful technology that is applied for genome editing of ncRNA functional characterization via loss of function or gain of function analysis (Pauli et al. 2014, 2015; Bazzini 2014). Currently, a new genome editing guide RNA nuclease was developed for the knockout in several ncRNAs. Clustered regularly interspaced short palindromic repeat (CRISPR)/ CRISPR-associated protein 9 (Cas9) system has been used to obtain potential and rapid genome editing in plant species (Jacobs et al. 2015). The big data domain is essential for genomic studies as it generates a massive volume of information when researchers sequence, analyze, and map genomes. Every data contains more than 20,000–25,000 genes; thus, it produces gigabytes of data comparable with lakhs of photograph data (Swathi et al. 2020). In this chapter, we describe the functional characters, classification, and novel technologies for the identification of ncRNAs in plants by big data analysis.

7.2 Different Kinds of ncRNAs in Plants

7.2.1 Circular RNA (circRNAs)

Discovery of many circRNAs over the range of plant species (Lee et al. 2017) has been demonstrated, for example, *Actinidia chinensis* (Wang et al. 2017c) and *Solanum lycopersicum* (Zuo et al. 2016). Although there is very little information about the functional characters of circRNAs, their function reported in mammals may assist as a primary direction for upcoming investigations on the functional characterization of plant circRNAs. It has been previously reported in the Arabidopsis plant that circRNAs originated from the exon-6 of the SEP3 (SEPALLATA3) gene which can bind with exact regions to its analogous DNA and produces R-loop (i.e., RNA:DNA hybrid). The formation of R-loop structure results in pausing of the transcriptional process, which in turn drives the floral homeotic phenotypes (Conn et al. 2017). CircRNAs play a significant role in a broad range of developmental and biological processes (Zhang et al. 2013b) and perform as miRNA sponges (Hansen et al. 2013; Jeck et al. 2014). Moreover, it was also demonstrated that circRNAs are involved in the regulation of pathogen invasive (Wang et al. 2017c) and chilling stress condition (Zuo et al. 2016).

7.2.2 Linear ncRNAs

7.2.2.1 Small Nucleolar RNAs (snoRNAs)

It acts as a guide RNA to direct methylation and pseudouridylation of complementary rRNA in pre-rRNA processing and modification, respectively. snoRNAs primarily guide other RNA chemical modifications, mainly snRNAs, tRNAs, and rRNAs (Lizarbe et al. 2017). Based on the common sequence motifs and structural features, snoRNA has two main classes, the H/ACA box snoRNAs, which are associated with pseudouridylation, and the C/D box snoRNAs, which are associated with methylation (Liang et al. 2019). The C/D box snoRNAs show two conserved motifs, the 3′-end D box (5′-CUGA-3′) and the 5′-end C box (5′-UGAUGA-3′) which is generally flanked by small inverted repeats (Chen et al. 2003). While in case of H/ACA box snoRNAs, in addition to the hinge region of H box (ANANNA) and a 3 nt ACA motif from the 3′-end of the molecule, which are secondary structure personified by hairpin–hinge–hairpin–tail, this is necessary for rRNAs pseudouridylation (Ganot et al. 1997).

7.2.2.2 Small ncRNAs

MicroRNAs (miRNAs)

In mRNA degradation and translational repression pathways, miRNAs are working as gene silencers in the modification of posttranscriptional level. These 18–24 nucleotide (nt) small endogenous noncoding RNAs originate from long self-complementary precursors. From 5′-end of the miRNAs, 2–8 nt portion is called the "seed" sequence (Bartel 2009). Seed sequence plays an important role in the binding and recognition of an accurate target site of mRNA. On the basis of genomic location, miRNAs are further classified as conserved and non-conserved miRNAs (Liu et al. 2017). Biosynthesis of miRNAs in the plants differs from that in animals since miRNA in plants is formed independently of DICER (class III RNAse enzyme). The synthesis of miRNAs in animals commences with the transcription of primary microRNA (pri-miRNA) which is cleaved into 60–120 nt lengthy precursor microRNA (pre-miRNA) with the help of a microprocessor complex that consists of DROSHA (class II ribonuclease) and DGCR8 or DiGeorge syndrome critical protein 8 (dsRNA binding protein; Ha and Kim 2014). However, miRNA biosynthesis involves the association of other proteins like elavl-like 1, mothers against DPP (decapentaplegic) homologue 4, p68/72 RNA helicase, and KH-type splice regulatory protein to microprocessor complex (Trabucchi et al. 2009; Blahna and Hata 2012). The pre-miRNA is then exported out in the cytoplasm where it further undergoes cleavage into miRNA-miRNA* duplex which takes place via DICER. The duplex miRNA is then loaded onto the RNA-induced silencing complex (RISC) leading to translational repression and transcript degradation. The mature plant miRNA is 2'-O-methylated at its 3′ termini (Sundaram 2019). Plant miRNAs are involved in the developmental procedure including seed germination, morphogenesis, root and shoot development, reproductive and vegetative phase change, and flowering to fruit initiation (Jung et al. 2009; Jones-Rhoades et al. 2006). Moreover, plant miRNAs participate in gene regulatory networks, regulating various biological processes like nutrition accumulation (Guo et al. 2013), stress response, metabolism, siRNA pathway feedback regulation, DNA methylation, protein degradation, epigenetic silencing, pathogen resistance, signal transduction, and maintaining genome integrity (Liu et al. 2017).

Small Interfering RNA (siRNA)

The siRNAs are among the most distributed small ncRNAs in eukaryotes (Shabalina and Koonin 2008). They originate from DCL (Dicer-like) catalyzed process of dsRNA (double-stranded RNA) precursors (Axtell et al. 2013). It is also known as silencing RNA, has 20–27 nt in length, and operates within the RNA interference (RNAi) pathway, thus protecting genomes against viruses and other parasites (Nozawa and Kinjo 2016). It interferes through the expression of specific genes with complementary nt sequences by the degradation of mRNA after transcription. The siRNAs are of two types, exogenous siRNAs (called exo-siRNAs), which are synthesized from exogenous RNAs, and endogenous siRNAs (called endo-siRNAs), which are synthesized from endogenous transcripts like pseudogenes, transposons, intergenic regions, and heterochromatic sequences (Ghildiyal and Zamore 2009). The exo-siRNAs have been thought to arise before endo-siRNAs. Endo-siRNAs are involved in transposon and repetitive elements silencing, whereas pseudogene-derived endo-siRNAs regulate their functional homologs. Endo-siRNAs have been also derived from mRNAs, which are made of long hairpins, suggesting them to be the ancestral state of miRNAs (Nozawa and Kinjo 2016). Plant siRNAs can be typified as natural antisense transcript (NAT) siRNAs, hc-siRNAs (heterochromatic siRNAs), and phased, secondary, siRNAs (phasiRNAs) including trans-acting siRNAs (tasiRNAs).

Recent studies reported in the plant revealed that NAT-siRNA participated in salt-stress responses while hc-siRNA is involved in response to fungal elicitors (Wang et al. 2017b). PhasiRNAs are synthesized from mRNA transformed into dsRNA by the help of RDR6 (RNA-dependent RNA polymerase) and prepared by DCL4 which is exemplified by the tasiRNAs of Arabidopsis (Vazquez et al. 2004). In many cases, the trans-acting (tasiRNAs) name is used for a few phasiRNAs because they can perform as miRNAs within a homology-dependent process, regulating mRNAs slicing of AGO1-dependent from the genes other than that of their source mRNA (Fei et al. 2013).

tRNA-Derived Small RNA (tsRNA)

tRNA-derived 18–40 nt fragments have been reported from many organisms, which are designated as tsRNAs (Raina and Ibba 2014). Earlier it was thought that tsRNAs were random degradation products of tRNA. Moreover, emerging data have been representing that tsRNAs are not only an endonuclease products but also an essential operative molecule which are involved in stress response, genome stability maintenance, cell proliferation, and intergenerational epigenetic inheritance (Zhu et al. 2017). Many types of tsRNAs are found like ~20 nt 5′ and 3′ small tRNA-derived fragments (tRFs) and ~30 nt 3′ and 5′ tRNA-derived halves (tRHs). Generally these are synthesized from either mature tRNAs or pre-tRNAs (tRNA precursor). 3′ U tRFs and 5′ leader tRFs occurred from pre-tRNAs during the tRNA maturation time (Lee et al. 2009; Haussecker et al. 2010). 5′ tRF and 3′ tRF products are generated from D-loop and T-loop cleavages of mature tRNAs, respectively (Haussecker et al. 2010). The 5′ tRHs and 3′ tRHs occur from the anticodon loop cleavages (Zhu et al. 2017).

PIWI-Associated RNAs (piRNAs)

These are dicer independent 25–30 nt small ncRNAs that are derived from long single-stranded RNA (ssRNA) precursor that carries several piRNAs in the tandem (Ghildiyal and Zamore 2009). The piRNAs bind with the Piwi clade of the Argonaute proteins to create RISC which then binds with its target RNA and cuts the target site. The piRNAs are also synthesized by cleaved transcripts which are mainly obtained from transposons through piRNA amplification (known to be "ping-pong" cycle). These RNAs are mostly antisense; however, a few exhibits sense orientation. piRNAs mainly showed their expression in germline cells to suppress the activity of transposons at transcriptional and posttranscriptional levels. It directly plays a significant role in the chromatin modification to repress transcription (Nozawa and Kinjo 2016).

7.2.2.3 Long Noncoding RNA (lncRNAs)

lncRNAs have been shown to play crucial functions in plants especially with respect to their roles in development and stress responses, particularly resistance to fungal infection. Many studies have demonstrated their role in root organogenesis, vernalization, male sterility induction (Ding et al. 2012), seedling photomorphogenesis, alternative splicing, phosphate homeostasis, development of fruit and flower, modulation of epigenetic regulations and chromatin loop dynamics, abiotic stress response, and their role as transposable element associated lncRNAs (TE-lncRNAs; Singh et al. 2018). Among the four types of lncRNAs naturally occurring are sense transcripts, intergenic lncRNAs, intronic lncRNAs, and naturally occurring antisense transcripts (lnc-NATs). The "sense transcripts" have low expression and coding potential, the "intergenic lncRNAs" occurring in the intergenic regions have regulatory role on the adjacent genes (Derrien et al. 2012). The intronic regions of transcripts are capable of generating the "intronic lncRNAs," and the "lnc-NATs" originate from the antisense transcript strand (Ma et al. 2013). A distinct class of lncRNAs has been found which are known as long intergenic noncoding RNAs (lincRNAs) and arise from transposons. Transposable element associated lincRNAs (TE-lincRNAs) are a type of lincRNAs which are shown to occur in different plants like maize, Arabidopsis, and rice (Wang et al. 2017a).

Varieties of methods are used for identification of lncRNAs. Variants of RNA-seq involve identification of both coding and noncoding RNAs. Deep sequencing is majorly used for transcript identification with expression of low levels. RNA immunoprecipitation sequencing (called RIP-Seq) uses a protein to pool down of RNA which is immune-precipitated using the antibody, thus enabling transcriptome broad study for the complexes of RNA protein (Köster and Meyer 2018). Strand-specific RNA sequence has been used to recognize lncRNAs in orange, tomato, maize, etc. (Dominic et al. 2013). Essentially, the NGS technology-based RNA sequencing is RNA isolation by various techniques with subsequent data analysis. Although microarray applications like tiling arrays have also been used for lncRNA identification, the requirement of prior knowledge of sequence limits the ability of this method in identification of novel lncRNAs (Rai et al. 2018). A primary criterion of >200 nt length is used for identification of the lncRNAs. Using additional multiple filters and high confidence, lncRNAs are identified and assembled from the sequencing data and deposited in databases. A few examples of such databases are TAIR (The Arabidopsis Information Resource), PLncDB (Plant long non-coding RNA database), and the GREENC (Green non-coding database). Yet, bio-sequencing technologies suffer from inevitable limitations and error-proneness. This along with the emergence of big data has led to the development of predictive tools for RNAs. These tools use machine learning techniques for training of models with known lncRNAs that could be subsequently used for identification of novel lncRNAs (Singh et al. 2018).

7.3 Tools for Identification of ncRNAs

For the identification of ncRNAs, both experimental and computational approaches can be applied. However, the computational approaches are preferred over experimental approaches since the latter is complex, time-consuming, and tiresome involving a lot of resources. Bioinformatics approaches and NGS technologies provide a great advantage in identification and characterization of ncRNAs in plants.

7.3.1 miRPlant

It identifies the novel miRNAs enabling visualization of novel structure of pre-miRNA and locates small RNA reads relative to the miRNA hairpin structure. This user-friendly interface tool predicts novel plant miRNA with great accuracy. It does not need any other third-party tools such as RNA secondary structure prediction tools (An et al. 2014). Available at: http://www.australianprostatecentre.org/research/software/mirplant.

7.3.2 sRNA Toolbox

It provides features like sRNA expression profiling and several downstream analysis tools in deep sequencing experiments (Puerta et al. 2019). The sRNA toolbox consists of four tools:

1. sRNAbench which allows the profiling of known miRNAs expression and also other types of sncRNAs and analysis, detects isomiRs, and creates numerous graphical summaries, thus predicting novel miRNAs both in animals and plants. Available at: https://arn.ugr.es/srnatoolbox/srnabench/.
2. sRNAde which detects differentially expressed sRNA by applying three methods, edgeR, DEseq, and noiseq. It also provides a consensus differential expression file along with the output from each method. Moreover, the tool also performs isomiR analysis and cluster analysis (heat maps). Available at: https://arn.ugr.es/srnatoolbox/srnade/.
3. sRNAblast which performs deep sequencing reads by blast analysis against NCBI databases. It aims at analyzing the reads which could not be mapped in sRNAbench or other sRNA toolbox tools. The results obtained from this analysis thus predict the unexpected contamination sources like bacterial or viral RNA molecules. Available at: https://arn.ugr.es/srnatoolbox/srnablast/.
4. miRNAconsTarget which predicts miRNA, aims on user designated sets of miRNAs and 3′-UTRs, and can be either animal consTarget or plantconsTarget depending upon the species under consideration. It uses a selection microRNA target prediction program and applies them to the user supplied sets of microRNAs and 3′-UTRs. It reports individual predictions as well as consensus predictions. Available at: https://bioinfo5.ugr.es/srnatoolbox.

7.3.3 miRA

It is a miRNA identification tool that is suitable for organisms with no existing annotation of miRNA or without a known relevant organism with properly characterized miRNAs. The tool has shown the capability to recognize species-specific miRNAs (Evers et al. 2015). Available at: https://github.com/mhuttner/miRA.

7.3.4 Semirna

Short for searching for miRNAs against nt sequences, this tool identifies miRNAs using target sequence. It can also identify miRNAs in newly sequenced genomes with high accuracy using homology (Muñoz-Mérida et al. 2012). Available at: http://www.bioinfocabd.upo.es/semirna.

7.3.5 Tapir

This tool predicts plant miRNA targets and target mimics, which were undetectable by traditional tools. Available at: http://bioinformatics.psb.ugent.be/beg/tools/tapir.

7.3.6 psRNATarget

It stands for plant regulatory small RNAs (sRNAs), which include miRNAs and phased siRNAs (phasiRNAs). It predicts miRNA targets, including target site accessibility evaluation and the reverse complementary matching. This tool is based on a parallel repetitive Smith-Waterman algorithm, allowing analysis of a wide range of genomes through the high-throughput sRNA-seq data. It also calculates the multiplicity of the target sites. It has a user-friendly interface that presents data clearly and concisely (Dai et al. 2018). Available at: http://plantgrn.noble.org/psRNATarget/.

7.3.7 MicroPC

MicroPC or μPC stands for predicting and comparing plant miRNAs. The tool involves identifying miRNAs and predicting their targets from large-scale expressed sequence tags (ESTs) analysis. It provides complete information about conservation, evolution, and function of plant miRNAs via incorporated utilities (Mhuanthong and Wichadakul 2009). Available at: http://www3a.biotec.or.th/micropc/index.html.

7.3.8 C-mii

This software is used for target prediction and identification of miRNAs (Numnark et al. 2012). Available at: http://www3a.biotec.or.th/c-mii/.

7.3.9 MTide

This tool is involved in the identification of miRNA and their target interaction by the associating modified algorithm of CleaveLand4 and miRDeep2 for better understanding of regulatory interaction networks (Zhang et al. 2015). Available at: http://bis.zju.edu.cn/MTide/.

7.3.10 BioVLAB-MMIA-NGS

This Cloud-based system allows integrated analysis of miRNAs and mRNAs (called miRNA mRNA Integrated Analysis or MMIA) using high-throughput sequencing data. MMIA is expected to reveal potential biological relationships between miRNA expression and their targeting mRNA expression changes, thus allowing genome-scale measurement of expression. The tool thus provides the results of gene set analysis, miRNA mRNA target network analysis, alignment visualization, pathway analysis, and novel miRNA prediction (Chae et al. 2015). Available at: http://epigenomics.snu.ac.kr/biovlab_mmia_ngs/.

7.3.11 PhyloCSF

It involves calculating lncRNA coding potential with the help of codon substitution frequencies or CSF score. It is a method that determines whether a multispecies nt sequence alignment is likely to represent a protein-coding region. Instead of relying on homology to known protein sequences, the tool examines the evolutionary signature characteristic to alignments of conserved coding regions. It also helps in distinguishing protein-coding and noncoding RNAs (Lin et al. 2011). Available at: https://github.com/mlin/PhyloCSF.

7.3.12 CPC

The full form of this tool is coding potential calculator that is a support vector machine (SVM)-based classifier to calculate lncRNA coding potential using six biologically meaningful sequence features. This user-friendly web-based tool can differentiate coding from noncoding transcripts and graphically display the detailed sequence features and additional annotations of the transcript with high accuracy (Kang et al. 2017). Available at: http://cpc.cbi.pku.edu.cn.

7.3.13 CNCI

Coding-Non-Coding Index or CNCI is a bioinformatics tool that calculates lncRNA coding potential, independent of known annotations, by profiling adjoining nt triplets. CNCI has been found to classify sense-antisense pairs and incomplete transcripts with high accuracy (Sun et al. 2013). Available at: https://github.com/www-bioinfo-org/CNCI#readme.

7.3.14 CPAT

Coding Potential Assessment Tool or CPAT is an alignment-free method that calculates the lncRNA coding potential using a logistic regression model built with four sequence features (ORF size, ORF coverage, Fickett TESTCODE statistic, and hexamer usage bias). It rapidly recognizes coding and noncoding transcripts. This tool outperforms CPC and PhyloCSF tools showing a specificity of 0.97 as opposed to 0.74 and 0.63, respectively. It is also faster than CPC and PhyloCSF (Wang et al. 2013). Available at: https://cpat.readthedocs.io/en/latest/.

7.3.15 Deeplnc

It is a deep neural network (DNN) classifier-based tool to predict lncRNAs. It is fast and accurate as compared to other existing classifiers (Tripathi et al. 2016). Available at: https://bioserver.iiita.ac.in/deeplnc/index.php.

7.3.16 spongeScan

This tool is designed for the identification and visualization of putative miRNA response elements (MREs) in lncRNA sequences. spongeScan has been found to have great potential in discovering lncRNA and miRNA cross talk (Furió-Tarí et al. 2016). Available at: http://spongescan.rc.ufl.edu/.

7.3.17 RegRNA

This tool is involved in the identification of regulatory RNA motifs and functional sites in an RNA sequence. This user-friendly tool allows graphical visualization of results (Huang et al. 2006). Available at: http://regrna2.mbc.nctu.edu.tw/index.html.

7.3.18 PredcircRNA

This tool identifies circRNA from other lncRNAs and introduces a computational approach from the transcript sequence assembled from RNAseq by using Cufflinks19, to differentiate circRNA from other lncRNAs. In this tool added are discriminative features from different sources of transcript sequences, like graphical feature, conservation information, SNP density, sequence composition, open reading frame (ORF), ALU, and tandem repeat (Pan and Xiong et al. 2015). Available at: https://github.com/xypan1232/PredcircRNA.

7.3.19 CircCode

This tool mainly focused on alignment between circRNA sequences and Ribo-seq reads which explore the locations of circRNAs and determine their coding potential by using mapping of the Ribo-seq reads. Also, it can enhance the accuracy rate and save the huge amount of computational time (Sun and Li 2019). Available at: https://github.com/PSSUN/CircCode.

7.3.20 CircPlant

It is an integrated tool for circRNA identification and functional characterization of plants, including circRNA-miRNA interactions, gene ontology (GO) annotation, and competing endogenous RNA pair (or circRNA-miRNA-mRNA networks) analysis (Zhang et al. 2020). Available at: http://bis.zju.edu.cn/circplant.

7.3.21 Circseq-Cup

This tool reported about 3000 complete lengths of rice circRNAs and also discovered that non-GT/AG rich splicing signals are widespread in the rice circRNAs (Ye et al. 2017). Available at: https://github.com/bioinplant/circseq-cup/.

7.3.22 PcircRNA_finder

This circRNA detecting tool provided more accurate and precise prediction process than other commonly used programs (like CIRCexplorer and find_circ) for rice and *Arabidopsis thaliana* (Chen et al. 2016) Available at: http://ibi.zju.edu.cn/bioinplant/tools/manual.htm.

7.3.23 AsmiR

This tool identifies the target sites of miRNA situated in alternately spliced regions of circular and linear splice isoforms via the current SMRT (single-molecule real-time), Illumina sequencing, and Iso-Seq (isoform sequencing) data in 11 plant species. This tool also can provide unique relationships between alternative splicing (AS) and miRNAs of posttranscriptional regulation in both gymnosperms and angiosperms (Wang et al. 2019a). Available at: http://forestry.fafu.edu.cn/bioinfor/db/ASmiR.

7.3.24 miRkwood

It is a very competent and user-friendly tool for the identification and annotation of plant miRNA genomes (Guigon et al. 2019). Available at: http://bioinfo.cristal.univ-lille.fr/mirkwood.

7.3.25 Si-Fi

It is open-source software also known as siRNA-Finder (Lück et al. 2019). It has efficiency for prediction of RNAi sequences, specificity, and design optimization capacity which is required for off-target search and practical application. Available at: https://www.genscript.com/tools/sirna-target-finder.

7.3.26 pssRNAit

This is a plant-specific RNAi tool, which can be used to design prediction of a pool siRNAs for more specific, effective, and nontoxic gene silencing in plants. This computational advance tool provided the transcript dataset of more than 160 plant species and also many other information like the RNAi pathway of the biological mechanism and several gene silencing mechanisms (Ahmed et al. 2020). Available at: https://plantgrn.noble.org/pssRNAit/.

7.4 Databases for ncRNAs

Big data refers to the inordinate structure and unstructured data and is outlined by 5 V's, namely, volume, velocity, variety, veracity, and value. Volume refers to the huge amount of data generated, velocity refers to the high speed at which the data is generated, variety is the type of data generated from different sources, veracity is the accuracy and trustworthiness of the generated data, and value is the benefit of perceiving the data (Swathi et al. 2020). However, big data poses a significant challenge in processing and analyzing using traditional methods. Thus, a novel machine learning tool by deep learning method provides an appropriate and accurate approach to big data analytics, and it has been implemented for this humongous amount of data (Nagaraj et al. 2018). Its application includes image and speech recognition, natural language processing, automatic navigation systems, and text mining. Deep learning has put down its roots in various domains including genomics, proteomics, transcriptomics, and structural biology and thus can be used for identifying novel RNAs in plants. Big data analytics have helped scientists to have a better look at genes and understand its related issues. Present time biologists employ big data analytics for genomic data rather than performing procedures in laboratories to identify biomarkers for new diseases. The databases storing data for ncRNAs are described as below.

7.4.1 miRBase

Managed by the researchers of University of Manchester and funded by Biotechnology and Biological Sciences Research Council, UK, this database is an archive of miRNA sequences and annotations giving information about their location and sequences. Recent release of miRBase (v22) version contains miRNA information of 48,860 mature miRNAs and 38,589 hairpin precursors belonging to 271 organisms (Kozomara et al. 2019). Available at: http://www.mirbase.org/.

7.4.2 Rfam

This database is a collection of RNA families' information, each of them presented by multiple alignments of sequence, covariance models (CMs), and consensus secondary structures based on homology relationship. It is broadly classified into three functional classes: noncoding RNA genes, self-splicing RNAs, and structured cis-regulatory elements. Although more complicated than Rfam (which uses hidden Markov models (HMMs)), the CMs in this database can simultaneously model RNA sequence and its structure more accurately (Griffiths-Jones et al. 2003). Available at: http://rfam.xfam.org/.

7.4.3 PmiRKB

Plant miRNA Knowledge Base or PmiRKB consists of miRNA of two model plants *A. thaliana* and *Oryza sativa*. This database provides four major functional modules. Available at: http://bis.zju.edu.cn/pmirkb/.

1. SNP: This module detects single nt proteins (SNPs) in pre-miRNAs, miRNAs, and target sites taking into reference the SNPs of seven accessions of Arabidopsis and two subspecies of rice.

2. Pri-miR: This module is used to investigate the transcriptional contexts and tissue-specific of pri-miRNAs and pre-miRNAs taking into reference the MPSS data developed from poly A-tailed transcripts.
3. MiR–Tar: From the parallel analysis of RNA end (PARE) data derived from degradomes, this tool was considered to authenticate thousands of miRNA pair target.
4. Self-reg: This module refers to data derived from degradomes to explore the precursors of miRNA metabolism including processing and self-regulation of miRNA-mediated effects on their host precursors.

7.4.4 PMRD

The plant miRNA database (PMRD) integrates the available plant microRNA data, consisting of miRNA sequence and their target genes, expression profiling, genome browser, and secondary dimension structure (Zhang et al. 2010). Available at: http://bioinformatics.cau.edu.cn/PMRD/.

7.4.5 PlanTE-MIR

This database delivers positional intersections between annotated transposable elements (TEs) and pre-miRNAs in plant genomes (Lorenzetti et al. 2016). Available at: http://bioinfo-tool.cp.utfpr.edu.br/plantemirdb/.

7.4.6 PlaNC-TE

This database consists of information about several classes of noncoding RNAs for the analysis of all plant species available at Ensembl Plants for determining the overlapping regions TE and ncRNAs (Pedro et al. 2018). Available at: http://planc-te.cp.utfpr.edu.br/.

7.4.7 miRTarBase

This experimentally validated miRNA-target interaction database is an inventory of over 360,000 MTIs (miRNA-target interactions), which are collected by manually surveying relevant literature after natural language processing of the manuscript to filter research articles related to functional characters studies of miRNAs. The accumulated MTIs are certified experimentally by microarray, western blot, reporter assay, and NGS experiments (Huang et al. 2020). Available at: http://mirtarbase.cuhk.edu.cn/php/index.php.

7.4.8 PASmiR

This is a literature-based database for miRNA molecular regulation in plant response to abiotic stress. This is a comprehensive repository which provides a solid platform for collection, searching, and standardization of miRNA stress regulation data in plants (Zhang et al. 2013a). Available at: http://pcsb.ahau.edu.cn:8080/PASmiR/f.

7.4.9 WMP

Wheat miRNA Portal or WMP is a comprehensive wheat miRNA database for functions of miRNAs in abiotic stress condition, thus providing direct access to online miRNA prediction software for de novo miRNA detection in wheat (Remita et al. 2016). Available at: http://wheat.bioinfo.uqam.ca/.

7.4.10 NONCODE

It is an integrated knowledge database for ncRNAs (excluding rRNAs and tRNAs) in 39 species consisting of 16 animals and 23 plants. It includes literature and other public databases searched from PubMed, Ensembl, RefSeq, lncRNAdb, and GENCODE. Besides providing the basic information of lncRNA such as location, exon number, strand, sequence, and length, it also provides information about the expression profile, predicted function, conservation info, exosome expression profile, and disease relation (Bu et al. 2011). Available at: http://www.noncode.org/.

7.4.11 GREENC

It is a plant lncRNA database based on wiki with more than 200,000 annotated transcripts (Gallart et al. 2016). Available at: http://greenc.sciencedesigners.com/wiki/Main_Page.

7.4.12 PNRD

PNRD or Plant Non-coding RNA Database is an updated version of PMRD. This database consists of many functional searches like miRNA-target search, literature-based function search, ncRNA keyword search, and online analysis like customized UCSC genome browser, new miRNA prediction toolkit, Blast tool, and coding potential calculator toolkit (Yi et al. 2015). Available at: http://structuralbiology.cau.edu.cn/PNRD/.

7.4.13 CANTATAdb

This database is a repository of 239,631 lncRNAs that are computationally recognized in 36 plant species and three algae. The database presents lncRNA sequences, genomic locations, and expression values across RNA-Seq libraries, noncoding RNAs from NONCODE, BLAST search results against Swiss-Prot proteins, and hypothetical peptides encoded by lncRNAs (Szcześniak et al. 2019). Available at: http://yeti.amu.edu.pl/CANTATA/.

7.4.14 Plant snoRNA Database

This plant-based snoRNA database provided the information on *Arabidopsis* and 18 other plant species including sequence data, initial gene orientation (polycistronic, intronic and single gene), pseudouridylation, and methylation target modification sites, expression data, and gene variant numbers (Brown et al. 2003). Available at: http://bioinf.scri.sari.ac.uk/cgi-bin/plant_snorna/home.

7.4.15 ASRG

This database recorded the 74 snRNA genes, *U4atac* snRNA gene, and 395 splicing-related genes of *Arabidopsis* along with their classification, gene structure, duplications, alternative splicing, and phylogenetic relationships (Wang et al. 2004). Available at: http://www.plantgdb.org/SRGD/WB04.php.

7.4.16 AtCircDB

This is a tissue-specific database of 622 RNA sequences of *Arabidopsis* which provided information of 84,685 circRNAs, 3486 microRNA-circRNA interactions, and 30,648 tissue-specific circRNAs (Ye et al. 2019). Available at: http://genome.sdau.edu.cn/circRNA.

7.4.17 PlantcircBase

: It provided total number of 77,595 distinctive circRNAs from five plant tissues (*O. sativa*, *Zea mays*, *Solanum lycopersicum*, *A. thaliana*, and *Hordeum vulgare*) including 1861 miRNA sponges and 1335 circRNA-miRNA-mRNA networks (Chu et al. 2017). Available at: http://ibi.zju.edu.cn/plantcircbase/.

7.4.18 PlantCircNet

PlantCircNet provided a comprehensible database for regulatory network analysis of circRNA–miRNA–mRNA, including interaction, visualization, information browsing, and database search (Zhang et al. 2017). Available at: http://bis.zju.edu.cn/plantcircnet/index.php.

7.4.19 CropCircDB

This database is incorporated with circRNA data sets which includes 38,785 from maize and 63,048 from rice (Wang et al. 2019b). Available at: http://genome.sdau.edu.cn/crop/.

7.5 Conclusion

The ncRNAs play crucial functions in a range of physiological processes. The rapid advancement in bioinformatics tools and high-throughput sequencing has provided innovatory ways for the identification and prediction of novel ncRNAs. However, a lot of information is yet to be discovered. Hence, there is a need to advance more on bioinformatics methods for the ncRNA identification, functional character investigation, and also construction of cross talk networks among various ncRNAs.

References

Ahmed F, Kumar MS, Dai X, Ramu VS, Lee S, Mysore KS, Zhao PX (2020) pssRNAit: a web server for designing effective and specific plant siRNAs with genome-wide off-target assessment. Plant Physiol 184(1):65–81. https://doi.org/10.1104/pp.20.00293

An J, Lai J, Sajjanhar A, Lehman ML, Nelson CC (2014) miRPlant: an integrated tool for identification of plant miRNA from RNA sequencing data. BMC Bioinformatics 15(1):275. https://doi.org/10.1186/1471-2105-15-275

Axtell MJ (2013) Classification and comparison of small RNAs from plants. Annu Rev Plant Biol 64:137–159. https://doi.org/10.1146/annurev-arplant-050312-120043

Bartel DP (2009) MicroRNAs: target recognition and regulatory functions. Cell 136:215–233. https://doi.org/10.1016/j.cell.2009.01.002

Bazzini AA, Johnstone TG, Christiano R et al (2014) Identification of small ORFs in vertebrates using ribosome foot printing and evolutionary conservation. EMBO J 33:981–993. https://doi.org/10.1002/embj.201488411

Blahna MT, Hata A (2012) Smad-mediated regulation of microRNA biosynthesis. FEBS Lett 586 (14):1906–1912. https://doi.org/10.1016/j.febslet.2012.01.041

Brown JSW, Echeverria M, Qu LH, Lowe TM, Bachellerie JP, Hüttenhofer A, Kastenmayer JP, Green PJ, Shaw P, Marshall DF (2003) Plant snoRNA database. Nucleic Acids Res 31 (1):432–435. https://doi.org/10.1093/nar/gkg009

Bu D, Yu K, Sun S, Xie C, Skogerbø G, Miao R, Xiao H, Liao Q, Luo H, Zhao G, Zhao H, Liu Z, Liu C, Chen R, Zhao Y (2011) NONCODE v3.0: integrative annotation of long noncoding RNAs. Nucleic Acids Res 40:D210–D215. https://doi.org/10.1093/nar/gkr1175

Chae H, Rhee S, Nephew KP, Kim S (2015) BioVLAB-MMIA-NGS: microRNA-mRNA integrated analysis using high-throughput sequencing data. Bioinformatics 31(2):265–267. https://doi.org/10.1093/bioinformatics/btu614

Chen CL, Liang D, Zhou H, Zhuo M, Chen YQ, Qu LH (2003) The high diversity of snoRNAs in plants: identication and comparative study of 120 snoRNA genes from Oryza sativa. Nucleic Acids Res 31(10):2601–2613. https://doi.org/10.1093/nar/gkg373

Chen L, Yu Y, Zhang X, Liu C, Ye C, Fan L (2016) PcircRNA_finder: a software for circRNA prediction in plants. Bioinformatics 32(22):3528–3529. https://doi.org/10.1093/bioinformatics/btw496

Chu Q, Zhang X, Zhu X, Liu C, Mao L, Ye C, Zhu QH, Fan L (2017) PlantcircBase: a database for plant circular RNAs. Mol Plant 10(8):1126–1128. https://doi.org/10.1016/j.molp.2017.03.003

Conn VM, Hugouvieux V, Nayak A, Conos SA, Capovilla G, Cildir G, Jourdain A, Tergaonkar V, Schmid M, Zubieta C, Conn SJ (2017) A circRNA from SEPALLATA3 regulates splicing of its cognate mRNA through R-loop formation. Nat Plants 3:17053. https://doi.org/10.1038/nplants.2017.53

Dai X, Zhuang Z, Zhao PX (2018) psRNATarget: a plant small RNA target analysis server. Nucleic Acids Res 46(W1):W49–W54. https://doi.org/10.1093/nar/gky316

Derrien T, Johnson R, Bussotti G, Tanzer A, Djebali S, Tilgner H et al (2012) The GENCODE v7 catalog of human long noncoding RNAs: analysis of their gene structure, evolution, and expression. Genome Res 22(9):1775–1789. https://doi.org/10.1101/gr.132159.111

Ding J, Lu Q, Ouyang Y, Mao H, Zhang P, Yao J, Yao J, Xu C, Li X, Xiao J, Zhang Q (2012) A long noncoding RNA regulates photoperiod-sensitive male sterility, an essential component of hybrid rice. Proc Natl Acad Sci 109:2654–2659. https://doi.org/10.1073/pnas.1121374109

Dominic Mills J, Kawahara Y, Janitz M (2013) Strand-specific RNASeq provides greater resolution of transcriptome profiling. Curr Genomics 14:173–181. https://doi.org/10.2174/1389202911314030003

Evers M, Huttner M, Dueck A, Meister G, Engelmann JC (2015) miRA: adaptable novel miRNA identification in plants using small RNA sequencing data. BMC Bioinformatics 16:370. https://doi.org/10.1186/s12859-015-0798-3

Fei Q, Xia R, Meyers BC (2013) Phased, secondary, small interfering RNAs in posttranscriptional regulatory networks. Plant Cell 25(7):2400–2415. https://doi.org/10.1105/tpc.113.114652

Furió-Tarí P, Tarazona S, Gabaldón T, Enright AJ, Conesa A (2016) spongeScan: a web for detecting microRNA binding elements in lncRNA sequences. Nucleic Acids Res 44(W1): W176–W180. https://doi.org/10.1093/nar/gkw443

Gallart AP, Pulido AH, de Lagrán IAM, Walter S, Cigliano RA (2016) GREENC: a wiki-based database of plant lncRNAs. Nucleic Acids Res 44(D1):D1161–D1166. https://doi.org/10.1093/nar/gkv1215

Ganot P, Bortolin ML, Kiss T (1997) Site-specific pseudouridine formation in preribosomal RNA is guided by small nucleolar RNAs. Cell 89:799–809. https://doi.org/10.1016/S0092-8674(00)80263-9

Ghildiyal M, Zamore D (2009) Small silencing RNAs: Anexpanding universe. Nat Rev Genet 10: 94–108. https://doi.org/10.1038/nrg2504

Griffiths-Jones S, Bateman A, Marshall M, Khanna A, Eddy SR (2003) Rfam: an RNA family database. Nucleic Acids Res 31(1):439–441. https://doi.org/10.1093/nar/gkg006

Guigon I, Legrand S, Berthelot JF, Bini S, Lanselle D, Benmounah M, Touzet H (2019) miRkwood: a tool for the reliable identification of microRNAs in plant genomes. BMC Genomics 20:532. https://doi.org/10.1186/s12864-019-5913-9

Guo S, Xu Y, Liu H, Mao Z, Zhang C, Ma Y, Zhang Q, Meng Z, Chong K (2013) The interaction between OsMADS57 and OsTB1 modulates rice tillering via DWARF14. Nat Commun 4:1566. https://doi.org/10.1038/ncomms2542

Ha M, Kim VN (2014) Regulation of microRNA biogenesis. Nat Rev Mol Cell Biol 15 (8):509–524. https://doi.org/10.1038/nrm3838

Hansen TB, Jensen TI, Clausen BH, Bramsen JB, Finsen B, Damgaard CK, Kjems J (2013) Natural RNA circles function as efficient microRNA sponges. Nature 495:384–388. https://doi.org/10.1038/nature11993

Haussecker D, Huang Y, Lau A, Parameswaran P, Fire AZ, Kay MA (2010) Human tRNA-derived small RNAs in the global regulation of RNA silencing. RNA 16:673–695. https://doi.org/10.1261/rna.2000810

Holley RW, Apgar J, Everett GA et al (1965) Structure of a ribonucleic acid. Science 147:1462–1465. https://doi.org/10.1126/science.147.3664.1462

Huang HY, Chien CH, Jen KH, Huang HD (2006) RegRNA: an integrated web server for identifying regulatory RNA motifs and elements. Nucleic Acids Res 34:W429–W434. https://doi.org/10.1093/nar/gkl333

Huang HY, Lin Y-C-D, Li J, Huang K-Y, Shrestha S, Hong H-C, Tang Y, Chen Y-G, Jin C-N, Yu Y, Xu J-T, Li Y-M, Cai X-X, Zhou Z-Y et al (2020) miRTarBase 2020: updates to the experimentally validated microRNA–target interaction database. Nucleic Acids Res 48(D1): D148–D154. https://doi.org/10.1093/nar/gkz896

Jacobs TB, LaFayette PR, Schmitz RJ, Parrott WA (2015) Targeted genome modifications in soybean with CRISPR/Cas9. BMC Biotechnol 15:16. https://doi.org/10.1186/s12896-015-0131-2

Jeck WR, Sharpless NE (2014) Detecting and characterizing circular RNAs. Nat Biotechnol 32 (5):453–461. https://doi.org/10.1038/nbt.2890

Jones-Rhoades MW, Bartel DP, Bartel B (2006) MicroRNAs and their regulatory roles in plants. Annu Rev Plant Biol 57:19–53. https://doi.org/10.1146/annurev-arplant-050718-100334

Jung JH, Seo PJ, Park CM (2009) MicroRNA biogenesis and function in higher plants. Plant Biotechnol Rep 3:111–126. https://doi.org/10.1007/s11816-009-0085-8

Kang YJ, Yang D, Kong L, Hou M, Meng YQ, Wei L, Gao G (2017) CPC2: a fast and accurate coding potential calculator based on sequence intrinsic features. Nucleic Acids Res 45(W1): W12–W16. https://doi.org/10.1093/nar/gkx428

Köster T, Meyer K (2018) Plant ribonomics: proteins in search of RNA partners. Trends Plant Sci 23(4):352–365. https://doi.org/10.1016/j.tplants.2018.01.004

Kozomara A, Birgaoanu M, Griffiths-Jones S (2019) miRBase: from microRNA sequences to function. Nucleic Acids Res 47(D1):D155–D162. https://doi.org/10.1093/nar/gky1141

Lee SM, Kong HG, Ryu CM (2017) Are circular RNAs new kids on the block? Trends Plant Sci 22: 357–360. https://doi.org/10.1016/j.tplants.2017.03.007

Lee YS, Shibata Y, Malhotra A, Dutta A (2009) A novel class of small RNAs: tRNA-derived RNA fragments (tRFs). Genes Dev 23:2639–2649. https://doi.org/10.1101/gad.1837609

Li J, Reichel M, Li Y, Millar AA (2014) The functional scope of plant microRNA-mediated silencing. Trends Plant Sci 19:750–756. https://doi.org/10.1016/j.tplants.2014.08.006

Liang J, Wen J, Huang Z, Chen X, Zhang B, Chu L (2019) Small nucleolar RNAs: insight into their function in cancer. Front Oncol 9:587. https://doi.org/10.3389/fonc.2019.00587

Lin MF, Jungreis I, Kellis M (2011) PhyloCSF: a comparative genomics method to distinguish protein coding and non-coding regions. Bioinformatics 27(13):i275–i282. https://doi.org/10.1093/bioinformatics/btr209

Liu D, Mewalal R, Hu R, Tuskan GA, Yang X (2017) New technologies accelerate the exploration of non-coding RNAs in horticultural plants. Hortic Res 4:17031. https://doi.org/10.1038/hortres.2017.31

Liu J, Jung C, Xu J et al (2012) Genome-wide analysis uncovers regulation of long intergenic noncoding RNAs in Arabidopsis. Plant Cell 24:4333–4345. https://doi.org/10.1105/tpc.112.102855

Liu J, Wang H, Chua NH (2015) Long noncoding RNA transcriptome of plants. Plant Biotechnol J 13:319–328. https://doi.org/10.1111/pbi.12336

Lizarbe MA, Espinosa JC, Lizarbe EF, Lizarbe SF, Robles MA, Olmo N, Turnay J (2017) Colorectal cancer: from the genetic model to posttranscriptional regulation by noncoding RNAs. Biomed Res Int 2017:38. https://doi.org/10.1155/2017/7354260

Lorenzetti APR, de Antonio AGY, Paschoal AR, Domingues DS (2016) PlanTE-MIR DB: a database for transposable element-related microRNAs in plant genomes. Funct Integr Genomics 16(3):235–242. https://doi.org/10.1007/s10142-016-0480-5

Lück S, Kreszies T, Strickert M, Schweizer P, Kuhlmann M, Douchkov D (2019) siRNA-finder (si-fi) software for RNAi-target design and off-target prediction. Front. Plant Sci 10:1023. https://doi.org/10.3389/fpls.2019.01023

Ma L, Bajic VB, Zhang Z (2013) On the classification of long noncoding RNAs. RNA Biol 10:925–933. https://doi.org/10.4161/rna.24604

Matsui A, Nguyen AH, Nakaminami K, Seki M (2013) Arabidopsis non-coding RNA regulation in abiotic stress responses. Int J Mol Sci 14:22642–22654. https://doi.org/10.3390/ijms141122642

Mhuanthong W, Wichadakul D (2009) Micro PC (μPC): a comprehensive resource for predicting and comparing plant microRNAs. BMC Genomics 10:366. https://doi.org/10.1186/1471-2164-10-366

Muñoz-Mérida A, Perkins JR, Viguera E, Thode G, Bejarano ER, Pérez-Pulido AJ (2012) Semirna: searching for plant miRNAs using target sequences. OMICS 16(4):168–177. https://doi.org/10.1089/omi.2011.0115

Nagaraj K, Sharvani GS, Sridhar A (2018) Emerging trend of big data analytics in bioinformatics: a literature review. Int J Bioinforma Res Appl 14(1-2):144–205. https://doi.org/10.1504/IJBRA.2018.089175

Nozawa M, Kinjo S (2016) Noncoding RNAs, origin and evolution of. Encycl Evolution Biol 3: 130–135. https://doi.org/10.1016/b978-0-12-800049-6.00181-5

Numnark S, Mhuantong W, Ingsriswang S, Wichadakul D (2012) C-mii: a tool for plant miRNA and target identification. BMC Genomics 13:S16. https://doi.org/10.1186/1471-2164-13-S7-S16

Pan X, Xiong K (2015) Predcirc RNA: computational classification of circular RNA from other long non-coding RNA using hybrid features. Mol Biosyst 11:2219–2226. https://doi.org/10.1039/c5mb00214a

Pauli A, Norris ML, Valen E et al (2014) Toddler: an embryonic signal that promotes cell movement via Apelin receptors. Science 343:1248636. https://doi.org/10.1126/science.1248636

Pauli A, Valen E, Schier AF (2015) Identifying (non-)coding RNAs and small peptides: challenges and opportunities. Bioessays 37:103–112. https://doi.org/10.1002/bies.201400103

Pedro DLF, Lorenzetti AP, Domingues DS, Paschoal AR (2018) Pla NC-TE: a comprehensive knowledgebase of non-coding RNAs and transposable elements in plants. Database (Oxford) 2018:1–7. https://doi.org/10.1093/database/bay078

Puerta EA, Lebrón R, Rueda A, Gómez-Martín C, Giannoukakos S, Jaspez D, Medina JM, Zubkovic A, Jurak I, Fromm B, Marchal JA, Oliver J, Hackenberg M (2019) sRNAbench and sRNAtoolbox 2019: intuitive fast small RNA profiling and differential expression. Nucleic Acids Res 47(W1):W530–W535. https://doi.org/10.1093/nar/gkz415

Rai MI, Alam M, Lightfoot DA, Gurha P, Afzal AJ (2018) Classification and experimental identification of plant long non-coding RNAs. Genomics 111(5):997–1005. https://doi.org/10.1016/j.ygeno.2018.04.014

Raina M, Ibba M (2014) tRNAs as regulators of biological processes. Front Genet 5:171. https://doi.org/10.3389/fgene.2014.00171

Remita MA, Lord E, Agharbaoui Z, Leclercq M, Badawi MA, Sarhan F, Diallo AB (2016) A novel comprehensive wheat miRNA database, including related bioinformatics software. Curr Plant Biol 7:31–33. https://doi.org/10.1016/j.cpb.2016.10.003

Santosh B, Varshney A, Yadava PK (2014) Non-coding RNAs: biological functions and applications. Cell Biochem Funct 33:14–22. https://doi.org/10.1002/cbf.3079

Shabalina SA, Koonin EV (2008) Origins and evolution of eukaryotic RNA interference. Trends Ecol Evol 23:578–587. https://doi.org/10.1016/j.tree.2008.06.005

Singh PK, Ganguli S, Pal A (2018) Functions of long non-coding RNAs in plants: a riddle to explore. Nucleus 61:261–272. https://doi.org/10.1007/s13237-018-0249-4

Sun L, Luo H, Bu D, Zhao G, Yu K, Zhang C, Liu Y, Chen R, Zhao Y (2013) Utilizing sequence intrinsic composition to classify protein-coding and long non-coding transcripts. Nucleic Acids Res 41(17):e166. https://doi.org/10.1093/nar/gkt646

Sun P, Li G (2019) Circ code: a powerful tool for identifying circRNA coding ability. Front Genet 10:981. https://doi.org/10.3389/fgene.2019.00981

Sundaram GM (2019) Dietary non-coding RNAs from plants: fairy tale or treasure? Non-coding RNA Res 4(2):63–68. https://doi.org/10.1016/j.ncrna.2019.02.002

Sunkar R, Li YF, Jagadeeswaran G (2012) Functions of microRNAs in plant stress responses. Trends Plant Sci 17:196–203. https://doi.org/10.1016/j.tplants.2012.01.010

Swathi P, Jyothi S, Revathi A (2020) Long non-coding RNA for plants using big data analytics—a review. In: Jyothi S, Mamatha D, Satapathy S, Raju K, Favorskaya M (eds) Advances in computational and bio-engineering, CBE 2019. Learning and analytics in intelligent systems, vol 15. Springer, Cham, pp 233–247. https://doi.org/10.1007/978-3-030-46939-9_21

Szcześniak MW, Bryzghalov O, Ciomborowska-Basheer J, Makałowska I (2019) CANTATAdb 2.0: expanding the collection of plant long noncoding RNAs. Methods Mol Biol 1933:415–429. https://doi.org/10.1007/978-1-4939-9045-0_26

Trabucchi M, Briata P, Garcia-Mayoral M, Haase AD, Filipowicz W, Ramos A, Gherzi R, Rosenfeld MG (2009) The RNA-binding protein KSRP promotes the biogenesis of a subset of microRNAs. Nature 459(7249):1010–1014. https://doi.org/10.1038/nature08025

Tripathi R, Patel S, Kumari V, Chakraborty P, Varadwaj PK (2016) Deep LNC, a long non-coding RNA prediction tool using deep neural network. Netw Model Anal Health Inform Bioinfo 5:21. https://doi.org/10.1007/s13721-016-0129-2

Vazquez F, Vaucheret H, Rajagopalan R, Lepers C, Gasciolli V, Mallory AC, Hilbert JL, Bartel DP, Crété P (2004) Endogenous trans-acting siRNAs regulate the accumulation of Arabidopsis mRNAs. Mol Cell 16:69–79. https://doi.org/10.1016/j.molcel.2004.09.028

Wang BB, Brendel V (2004) The ASRG database: identification and survey of *Arabidopsis thaliana* genes involved in pre-mRNA splicing. Genome Biol 5:R102. https://doi.org/10.1186/gb-2004-5-12-r102

Wang D, Qu Z, Yang L, Zhang Q, Liu ZH, Do T, Adelson DL, Wang ZY, Searle I, Zhu JK (2017a) Transposable elements (TEs) contribute to stress-related long intergenic noncoding RNAs in plants. Plant J 90:133–146. https://doi.org/10.1111/tpj.13481

Wang H, Wang H, Zhang H, Liu S, Wang Y, Gao Y, Xi F, Zhao L, Liu B, Reddy ASN, Lin C, Gu L (2019a) The interplay between microRNA and alternative splicing of linear and circular RNAs in eleven plant species. Bioinformatics 35(17):3119–3126. https://doi.org/10.1093/bioinformatics/btz038

Wang J, Meng X, Dobrovolskaya OB, Orlov YL, Chen M (2017b) Non-coding RNAs and their roles in stress response in plants. Genomics Proteomics Bioinformatics 15(5):301–312. https://doi.org/10.1016/j.gpb.2017.01.007

Wang K, Wang C, Guo B, Song K, Shi C, Jiang X, Wang K, Tan Y, Wang L, Wang L, Li J, Li Y, Cai Y, Zhao H, Sun X (2019b) CropCircDB: a comprehensive circular RNA resource for crops in response to abiotic stress. Database 2019:baz053. https://doi.org/10.1093/database/baz053

Wang L, Park HJ, Dasari S, Wang S, Kocher JP, Li W (2013) CPAT: coding-potential assessment tool using an alignment-free logistic regression model. Nucleic Acids Res 41(6):e74. https://doi.org/10.1093/nar/gkt006

Wang Z, Liu Y, Li D, Li L, Zhang Q, Wang S, Huang H (2017c) Identification of circular RNAs in kiwifruit and their species-specific response to bacterial canker pathogen invasion. Front Plant Sci 8:413. https://doi.org/10.3389/fpls.2017.00413

Wei J, Huang K, Yang C, Kang C (2017) Non-coding RNAs as regulators in epigenetics (review). Oncol Rep 37:3–9. https://doi.org/10.3892/or.2016.5236

Ye CY, Chen L, Liu C, Zhu QH, Fan L (2015) Widespread noncoding circular RNAs in plants. New Phytol 208:88–95. https://doi.org/10.1111/nph.13585

Ye CY, Zhang X, Chu Q, Liu C, Yu Y, Jiang W, Zhu QH, Fan L, Guoe L (2017) Full-length sequence assembly reveals circular RNAs with diverse non-GT/AG splicing signals in rice. RNA Biol 14(8):1055–1063. https://doi.org/10.1080/15476286.2016.1245268

Ye J, Wang L, Li S, Zhang Q, Zhang Q, Tang W, Wang K, Song K, Sablok G, Sun X, Zhao H (2019) AtCircDB: a tissue-specific database for Arabidopsis circular RNAs. Brief Bioinform 20 (1):58–65. https://doi.org/10.1093/bib/bbx089

Yi X, Zhang Z, Ling Y, Xu W, Su Z (2015) PNRD: a plant non-coding RNA database. Nucleic Acids Res 43(D1):D982–D989. https://doi.org/10.1093/nar/gku1162

Zhang P, Liu Y, Chen H, Meng X, Xue J, Chen K, Chen M (2020) CircPlant: An integrated tool for CircRNA detection and functional prediction in plants. Genomics Proteomics Bioinformatics 18 (3:352–335. https://doi.org/10.1016/j.gpb.2020.10.001

Zhang P, Meng X, Chen H, Liu Y, Xue J, Zhou Y, Chen M (2017) PlantCircNet: a database for plant circRNA–miRNA–mRNA regulatory networks. Database 2017:bax089. https://doi.org/10.1093/database/bax089

Zhang S, Yue Y, Sheng LWY, Fan G, Li A, Hu X, Shangguan M, Wei C (2013a) PASmiR: a literature-curated database for miRNA molecular regulation in plant response to abiotic stress. BMC Plant Biol 13:33. https://doi.org/10.1186/1471-2229-13-33

Zhang Y, Zhang XO, Chen T, Xiang JF, Yin QF, Xing YH, Zhu S, Yang L, Chen LL (2013b) Circular intronic long noncoding RNAs. Mol Cell 51(6):792–806. https://doi.org/10.1016/j.molcel.2013.08.017

Zhang Z, Jiang L, Wang J, Gu P, Chen M (2015) MTide: an integrated tool for the identification of miRNA-target interaction in plants. Bioinformatics 31(2):290–291. https://doi.org/10.1093/bioinformatics/btu633

Zhang Z, Yu J, Li D, Zhang Z, Liu F, Zhou X, Wang T, Ling Y, Su Z (2010) PMRD: plant microRNA database. Nucleic Acids Res 38(1):D806–D813. https://doi.org/10.1093/nar/gkp818

Zhu L, Ow DW, Dong Z (2017) Transfer RNA-derived small RNAs in plants. Sci China Life Sci 61 (2):155–161. https://doi.org/10.1007/s11427-017-9167-5

Zuo J, Wang Q, Zhu B, Luo Y, Gao L (2016) Deciphering the roles of circRNAs on chilling injury in tomato. Biochem Biophys Res Commun 479(2):132–138. https://doi.org/10.1016/j.bbrc.2016.07.032

Artificial Intelligence in Biomedical Image Processing

8

Jubie Selvaraj, Thangavelu Prabha, T. Durai Ananda Kumar, and Senthilkumar Palaniappan

Abstract

Medical images are essential for the health care disciplines such as surgical procedures, cardiology, ophthalmology, obstetrics and gynecology, dermatology, etc. Medical images are often used in radiological procedures in diagnostic, clinical research, and therapy management. The medical image processing technology is indeed very essential, as it allows the enhanced and improved performance of the medical image analysis. Medical imaging is a procedure used to attain images of the body parts in order to identify or study diseases. Medical imaging is developing rapidly due to developments in image processing techniques including, image recognition, analysis, and enhancement. Computer algorithms are utilized in image processing and pattern recognition, which enhances the visualization of image components. These algorithms mimic the whole system made of many other suggested processing elements in a configuration specified by a graphic schematic design. Artificial intelligence (AI) is extensively applied for the augmentation of medical imaging (processing and interpretation) procedures. This chapter summarizes the application of AI on

J. Selvaraj
Department of Pharmaceutical Chemistry, JSS College of Pharmacy, JSS Academy of Higher Education and Research, Ooty, Tamil Nadu, India

T. Prabha
Department of Pharmaceutical Chemistry, Nandha College of Pharmacy, Erode, Tamil Nadu, India

T. D. A. Kumar
Department of Pharmaceutical Chemistry, JSS College of Pharmacy, JSS Academy of Higher Education and Research, Mysuru, Karnataka, India

S. Palaniappan (✉)
Faculty of Pharmacy, Karpagam Academy of Higher Education, Coimbatore, Tamil Nadu, India
e-mail: drsenthilkumar.p@kahedu.edu.in

S. Singh (ed.), *Machine Learning and Systems Biology in Genomics and Health*,
https://doi.org/10.1007/978-981-16-5993-5_8

both simple and sophisticated image analysis techniques, different medical image processing tools, and different image processing algorithms in several aspects. The primary goal is to collect and analyze data to advise the users of various operating systems on which medical imaging software to be used while analyzing different kinds of images.

8.1 Introduction

A physical representation and visual perception of any objective is known as image. It depicts the object both in two-dimensional (2D) and three-dimensional (3D) data with color variation. These images are broadly grouped into analog and digital images.

- **Analog image:** Images representing the continuous variation property are known as analog images.
- **Digital image:** Image composed of several picture elements (called pixels) and image elements, which are captured from analog images.

8.1.1 Medical Images

Medical images provide details of the biological systems and became essential for the diagnosis and treatment strategies (Wallyn et al. 2019). Traditionally microscopic slide visualization efficiently differentiates the morphology of normal and diseased tissue and is used for pathological interpretations. However, these details are not adequate to provide subcellular level disease mechanism, hence scientists developed medical imaging system. Medical imaging offers organ function information and visualization of tissues, helping in the detection of disease states and revolutionized the field of medicine (Hosny et al. 2018). Medical images are essential for the health care disciplines such as surgical procedures, cardiology, ophthalmology, obstetrics and gynecology, dermatology, etc. These techniques are non-invasive and do not require procedures such as biopsy. Medical images include the following subcategories.

8.1.1.1 Radiological Images

The interaction of specific electromagnetic waves with the biological tissues captures the image.

(a) **Gamma Rays Imaging:** Gamma rays emitting isotopes capture the images of infectious tissues, tumor cells, bones, and tissues of various organs and provide the basis for the diagnosis of diseases (Donya et al. 2014). This branch is known as diagnostic nuclear medicine.

(b) **Positron Emission Tomography (PET):** It is useful in imaging the biologically active molecules embedded with the positron emission tracer, which quantifies the physiological functions (Fukuchi et al. 2017).

(c) **Infrared Imaging:** Infrared domain of the electromagnetic spectrum provides high-resolution and deep-resolved imaging of tissues, in particular, deeper layers of the skin (Dang et al. 2019).

(d) **X-Ray Radiography Images:** The anatomy of the biological system (bones) can be visualized through X-ray images.

1. *X-Ray fluoroscopy* provides a sequence of radiographic images at a particular frame rate, which is useful in invasive procedures of angiograms and biopsies (Balter et al. 2010).
2. *Computed tomography (CT) images*: It captures the cross-sectional images of the organs, bones, and blood vessels and provides slice-by-slice inspection (Computed tomography (CT), US-Food and Drug Administration (US-FDA) 2021).
3. *Mammography*: A two-dimensional (2D) image produced with the help of low energy X-rays (20–32 kVp). It provides details about masses, calcification about the breast tissue, and hence useful in the detection of breast cancer (Iranmakani et al. 2020).

(e) **Magnetic Resonance Imaging (MRI):** It provides three-dimensional (3D) images of anatomical structures based on the rotational axis of biological fluids. It serves for diagnosis and therapy monitoring. Position emission tomography (PET) combined with MRI enhances the clarity of the image details (Pichler et al. 2008).

(f) **Optical Coherence Tomography (OCT):** It measures the echo delay time of cross-sectional images and tissue back-scattered light to generate three-dimensional (3D) images. OCT is mainly useful in the diagnosis of retinal disease (Rodríguez et al. 2018).

(g) **Ultrasonography Images:** Ultrasound waves utilize mechanical sound waves for the generation of images. It is most useful in screening and diagnosing prenatal screening (Carovac et al. 2011).

8.1.2 Image Recognition vs. Image Processing

Image recognition and image processing differ from each other (not synonyms).

(a) **Image Recognition:** The ability of the system to distinguish between the objectives based on certain features is known as image recognition (DiCarlo et al. 2012). Image recognition is a field of computer vision that interprets images to aid in decision making. Computer vision utilizes deep learning to perform image processing, image classification, object detection, object segmentation, image coloring, image reconstruction, and image synthesis. It provides a higher level of understanding for digital images. Image recognition is the final stage of image processing.

(b) **Image Processing:** Image processing involves image manipulation in order to extract the specific image feature, which enhances the information extraction (Latif et al. 2019). Image processing is of two types.

1. ***Analog image processing*** is a manual processing program that may be used to edit actual pictures, prints, and other hard copies of pictures.
2. ***Digital image processing*** uses computer algorithms to manipulate digital pictures.

8.1.3 Computer-Assisted Image Processing

Computer algorithms are utilized in image processing and pattern recognition, which enhances the visualization of image components, whereas a naked eye cannot identify this information (Shamir et al. 2010). The patterns are connected to normal and abnormal tissue developments and related to diseases. In the case of tumor analysis, the diameter, volume, and vasculature of tissues are studied.

- Step 1: *Gathering and organizing data*. Computers perceive an image as a vector or raster image and subject them for analysis. The quality of data is essential and decides the recognition model.
- Step 2: *Building a predictive model*. The first stage is the training classification algorithm, which helps in distinguishing the patterns in images.

8.1.4 Medical Image Processing

Radiologists and pathologists evaluate the medical images for the purpose of diagnosis and to decide the treatment strategy. The clarity of interpretation influences the clinical decisions and outcomes. Image processing converts 2D images into 3D images using reconstruction and modeling techniques. Medical image processing enhances the display of the medical image for accurate analysis and conclusions (Deserno Né Lehmann et al. 2013).

Digital medical images can be processed and distributed through communication networks such as Picture Archiving and Communication Systems (PACS) and Digital Imaging and Communication in Medicine (DICOM) (Ratib et al. 1994).

(a) **Picture Archiving and Communication System (PACS):** It is the computerized management system for storage, transmission, retrieval, and display of medical images. PACS avoids data redundancy, enhances the clinical decision-making process, and provides automatic grouping/clustering of the medical images.
(b) **Digital Imaging and Communication in Medicine (DICOM):** It ensures the quality of the medical image management system.

8.1.4.1 Steps of Medical Image Processing

Following are the major phases of medical image processing.

(a) *Image Acquisition*: Scaling and color conversion
(b) *Image Enhancement*: Extracts hidden details
(c) *Image Visualization*: Manipulation of matrix
(d) *Image Restoration*: Process of recovering the damaged version of image
(e) *Image Segmentation*: Partitioning an image into constituent parts
(f) *Image Analysis*: Quantitative measurements and interpretations
(g) *Image Management*: Storage, communication, transmission, archiving, and retrieval of image data

8.1.5 Image Informatics and Medical Image Informatics

Conventionally, medical images were analyzed through signal analysis and statistical modeling (Belle et al. 2015). The increase in the size and dimensionality of medical data demands efficient methods. The advancements in computer hardware and software provide efficient handling of images incorporating information and communication technologies (ICT). The computer program involving image acquisition, retrieval, management, processing, analysis, and visualization for the purpose of deriving a better conclusion is known as image informatics. The cognitive interpretation of images is different from the computer-assisted interpretation of digital images with discrete pixels. Segmentation of medical images pool large datasets, characterization of these tissues provides insights on biomarkers and are the basis for the treatment. The majority of these data are unstructured; hence the data mining process is difficult. To overcome these limitations several repositories, namely, clinical data repositories (CDRs), electronic health and medical records (EHR/EMR), radiology and pathology archives, and clinical trial management systems (CTMS), were developed (Dash et al. 2019). These data repositories are very much helpful in drug discovery and related research. Earlier, the analysis of CT scan images required hours together. The advent of computer programs such as medical image analysis software made the data acquisition and reconstruction procedure very rapid and completes within seconds. National Cancer Institute SEER program initiated the digital pathology for the surveillance of cancer biology phenotype data (Duggan et al. 2016). FDA approval for the use of imaging software

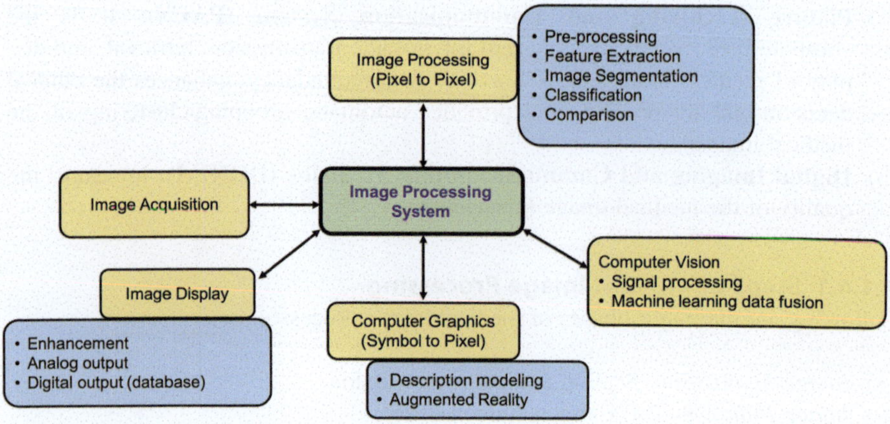

Fig. 8.1 Provision of digital image processing

(medical image informatics) was granted in the year 2017, which produced landslide changes in the clinical settings.

Medical image informatics, also known as radiology informatics, is widely applied in medical physics and extended for medical applications. It revolutionized the health care industry and drives the clinical setup interventions such as diagnosis, prognosis, and effective therapy planning. Medical image informatics assists in the generation, manipulation, management, and integration of medical images and is widely applied for improving the efficiency and reliability of the health care procedures such as physiology, pathophysiology, and clinical practice (Hsu et al. 2013). Society for Imaging Informatics in Medicine (SIIM) developed objectives to improve efficiency, accuracy, and reliability of medical image processing. SIIM insists on the effective exchange of images among health care professionals for the purpose of effective interpretations (Fig. 8.1).

8.1.6 Artificial Intelligence Assisted Image Processing

Diagnostic error is the significant factor for the mortality rate around the globe (~20 million per year). The major reason for the error is the inability of the radiologists particularly speaking of practicing radiologists in clinical reasoning. Artificial intelligence (AI) is extensively applied for the augmentation of medical imaging (processing and interpretation) procedures (Yoon et al. 2019). AI helps in transforming medical digital images into high-dimensional images (Barisoni et al. 2020), which are useful in clinical decision-making (diagnosis, prognosis). United States Food and Drug Administration (US-FDA) approval for AI-based techniques dominates the medical imaging informatics. US-FDA confirms that the use of AI tools in image search is harmless, suggests all the medical staff to implement AI and permits them to study the instrumentation. The number of AI algorithms were

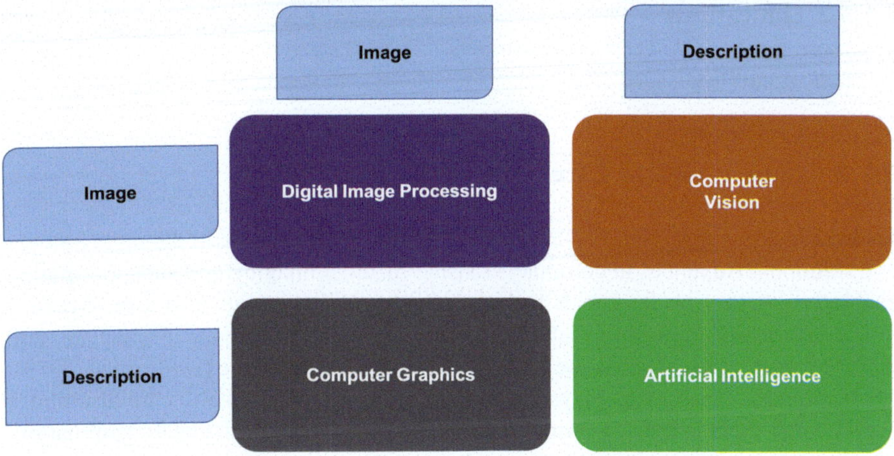

Fig. 8.2 Conventional vs Artificial Intelligence assisted Image processing

developed based on the training to identify the patterns (visible and invisible) from the medical images. This procedure mimics the human analytical cognition but offers enhanced detections and characterization based on sensitivity and specificity. This advanced feature is useful in handling larger populations and avoids the inaccuracy of the data. Machine learning (ML) and deep learning (DL) are applied for the improved performance of medical imaging procedures (Currie et al. 2019) (Fig. 8.2).

Unsupervised feature of AI permits high-throughput clinical data predictions such as disease detection, lesion segmentation, diagnosis and treatment selection including prognosis based on the images (Jiang et al. 2017). Deep learning-based automatic detection (DLAD) developed by a research group from Seoul National University and College of Medicine for the detection of abnormal cell growth outperformed 17 of 18 doctors (Nam et al. 2019). Machine learning detected tissue changes related to the ischemic stroke within a shorter duration. ***Image segmentation*** involves the decomposition of images into different patterns with an aid of artificial intelligence tool (Asokan et al. 2020). Face recognition features of number of electronic gadgets and Google Lens are few examples. Google Lens can process the image (decomposes the image into several parts) and identifies similar images from the database.

8.1.6.1 Advantages

- Image quality: It provides high quality images and interpretations, which are not amenable to human visualization.
- Time management: It decreases radiation dose and MRI scanner time, which is a major factor for the patient compliance.
- Accelerated decisions: AI escalate more effective conclusions and hence reduces the time required for the diagnosis.

8.1.6.2 Limitations

- Lower ability to detect motion artifacts
- Medical liability issues

8.1.7 Role in Health Care

Medical image informatics elevated the quality of health care with the most innovative solutions. Advances in computer-assisted medical imaging and supercomputing enable health care professionals to handle multiple data points at a time. Medical image informatics describes feature prognosis, insights of disease mechanism, and disease progression, which is the basis for personalized medicine (Uddin et al. 2019). Medical image informatics plays a significant role in several health care disciplines and become increasingly meaningful in preventing diseases, medical examination, deadly disease monitoring, patient care, diagnosis and treatment, disease progression analysis, treatment method selection, treatment effect assessment, and rehabilitation, in addition to disease diagnosis and treatment. Cutting-edge image processing aids doctors in properly interpreting medical pictures by providing superior tissue homogeneity, optimum contrast, edge enhancement, artifact reduction, intelligent noise control, and other features. The biomedical applications of medical image informatics are listed below.

- Medical image analysis delineates the images of interest and predicts the disease stage. Medical imaging informatics provides adequate knowledge for the diagnosis and treatment particularly in the case of cardiovascular and neurodegenerative diseases.
- Medical image informatics profiles a large pool of electronic patient health records (ePHRs) and enables the clinical decision with improved accuracy and hence delivers personalized treatment (personalized medicine).
- ML- and DL-based tools are made available to handle natural medical images. These tools are useful in the study of dermoscopic skin lesions, MRI images (brain and heart), disease models, and blood vessel segmentation through a number of algorithms.
- DL algorithms were reported for their potential for disease classification such as screening neurological diseases, diabetic retinopathy, and melanoma. Convolutional neural networks (CNN), a DL tool, was used for the differentiation of benign and malignant lung nodules using X-ray images.
- Radiogenomics, a branch emerged by the amalgamation of imaging and omics, helps in the construction of models for tissues to predict their functions (Zanfardino et al. 2019). This is effectively adopted for the visualization of cell proliferation, tumor expansion and metastasis.

Software-as-a-medical-device (SAMD) performs image analysis using machine learning and deep learning-based algorithms (Benjamens et al. 2020). The tools for image enhancement (e.g., SubtlePET/MR, IDx-DR), segmentation and diagnosis (e.g., Lung/LiverAI, OsteoDetect), and malignancy likelihood estimation (e.g., Transpara) are made available (Panayides et al. 2020).

8.2 Processes/Tools in Image Processing

The process of creating visual pictures of inside body systems for scientific and medical diagnosis and practice, as well as a visual view of the functioning of internal tissues, is known as medical imaging. The goal of this procedure is to identify and treat disorders. This procedure produces a database of the organ's normal structure and function, making it simple to distinguish between them. There are a variety of different technologies that may be employed to obtain data about the body's position and functionality. Biochemical and radiological imaging, using electromagnetic energy (X-rays and gamma rays), sonography, magnetic scopes, and thermal and isotope imaging are the part of this process. When compared to techniques that create pictures, such approaches have a lot of drawbacks. Every year, billions of pictures are taken throughout the world for various medical purposes. Ionizing and nonionizing radiation modifiers are used by almost half of them (Abdallah 2018).

Image processing techniques are a set of concepts for computer-assisted photo editing. The goal of segmentation is to divide the pictures into significant sections. Local segmentation is the division of pictures into tiny sections within a single image. The construction of the partitions is dealt with global segmentation. The three techniques of image segmentation processing are region, border, and edge. The region technique is used to look at pictures and determine the region class of adjacent pixels. The histogram and the threshold value of pixels are used in threshold segmentation. To evaluate pictures at boundaries or discontinuities, image edge methods are utilized.

AI in biomedical image processing is the usage of software and a complex structure of algorithms to mirror human intelligence in the analysis of composite medical data. The AI computer algorithms are used to estimate the results without direct human interaction (Davenport and Kalakota 2019). The key features of AI include patient-centric information retrieval, biomedical text mining, clinical event forecasting, data-driven prognostics and diagnostics, and precision medicine (Bohr and Memarzadeh 2020). Biomedical imaging through AI (machine learning) has been applied for studying brain activity, visualizing blood vessels, characterizing skin lesions, and cancer diagnosis. Nevertheless, the quality of medical images is dependent on the number of sensors distribution used by the apparatus.

Medical imaging creates pictures of the body's interior structures without the need for intrusive treatments. Such pictures were created with graphical interfaces and by converting energy to signals numerically and conceptually (Abdallah 2016). After that, the impulses are transformed into digital pictures. Those signals indicate many kinds of tissues that make up the human body.

8.2.1 Artificial Intelligence in Biomedical Visualization

8.2.1.1 Vision Recognition by Computer

In the health and biomedical industries, computer vision, also known as machine vision, is becoming more sophisticated and widely employed. Computer vision refers to a computer's capacity to identify, analyze, interpret, and comprehend visual characteristics. Computer vision can use texture, size, contour, and past knowledge, as well as contextual data from images, to aid in human visual comprehension. The usage of magnetic resonance imaging (MRI), computer tomography (CT), positron emission tomography (PET), and ultrasound imaging is prevalent in the medical sector, offering computer vision with a plethora of valuable data to extract. As a result, vision recognition by computer has become the ideal "tool" in twenty-first century medicine.

8.2.2 Deep Learning (DL) and Machine Learning (ML)

DL is a machine learning approach that builds artificial neural networks (ANNs) to imitate the structure and functioning of the brain. Because the machine's efficiency increases as it examines more data, this learning approach is a game-changing tool for processing massive amounts of data. DL is a simple solution to the challenge of assisting medical experts who are looking deeper into the data. DL has given a boost to the already rapidly expanding area of computer vision.

ML is transforming many components of the health care system, with image processing specialties like radiology and pathology. During each step, ML can help to enhance the radiology patient experience. ML might increase productivity from the adoption of smart workflow and reporting technologies. Besides picture processing, ML has numerous applications from assisting imaging divisions in maximizing their efficiency and productivity to a tool that enhances the image capture workflow, guaranteeing the greatest possible image for each inspection (Fig. 8.3). Machine learning techniques are involved in various fields of medical image processing such as triage, automated reporting, detection, quantification, decision support, prediction, differential diagnosis, and quality control.

8.2.3 Image Processing Algorithms

Computer algorithms play the most significant role in digital image processing such as digital image recognition, image reconstruction, image restoration, image enhancement, image data reduction and spectral image prediction. Image processing algorithms often are used for capturing the complete image and could be classified as low-level techniques, such as color improvement and noise removal; medium-level techniques, such as compression and binarization; and higher-level techniques, which include segmentation, sensing, and recognition algorithms that retrieve semantic information from the data gathered (Parker 2010).

Fig. 8.3 Machine learning in medical image processing

8.2.3.1 Image Processing via Morphological Algorithm

Morphological image processing attempts to eliminate defects from binary pictures generated by conventional thresholding. This method examines a picture using a tiny template called a structuring element, which is a small matrix with 0 and 1 values that are positioned on various potential positions in the picture and matched to the appropriate neighborhood pixels. It also aids in the flattening of images through the use of opening and closing procedures. Dilation and erosion are the two most common morphological image processing techniques. In an image, the dilation process adds pixels to the object boundary, whereas the erosion process eliminates pixels from the object boundary. The magnitude of the structuring element determines how many pixels are deleted or included in the original image.

8.2.3.2 Image Processing via Gaussian Algorithm

The consequence of blurring an image with a Gaussian function is known as Gaussian blur, often called Gaussian smoothing. It's a technique for reducing picture noise and distorting information. This blurring method has the same visual impact as looking at an image via a transparent screen. It's often used in machine learning for image improvement at various scales or as a deep learning data augmentation tool. In practice, it is also preferable to split the operation into two stages to make use of the

Gaussian blur's separable characteristic. A one-dimensional kernel is used in the first stage to blur the picture solely in the horizontal or vertical directions. The same one-dimensional kernel is utilized in the second stage to blurry the leftover directions. The outcome is the same as thresholding in a single stage using a two-dimensional kernel.

8.2.3.3 Image Processing Via Fourier Transform Algorithm

It is a very important tool, in which a picture is broken down into sine and cosine components using the Fourier transform algorithm. The picture in the Fourier or frequency domain is represented by the outcome of the transformation, whilst the spatial domain equivalent is represented by the input picture. Each spot in the Fourier domain picture indicates a frequency shown in the spatial domain picture. This algorithm may be used for picture processing, picture filtration, picture restoration, and image compression (Shi et al. 2017).

8.2.3.4 Image Processing via Edge Detection Algorithm

This algorithm is used to locate the edges of objects within pictures. It operates by sensing brightness inconsistencies. Since most of the shape features are included in the edges, this might be highly effective in extracting valuable information from the image. This could react quickly if it detects noise in the picture while analyzing fluctuations in grey levels. Edges are described as the gradient's local maxima. The Sobel edge detection method is the most widely used edge detection algorithm.

8.2.3.5 Image Processing via Wavelet Algorithm

Fourier transform image processing approach is restricted to frequency; nevertheless, wavelets incorporate both the time and frequency. For nonstationary signals, this transformation is appropriate. The edges are the crucial element of a picture but when using conventional filters, it has been observed that noise is eliminated; however, the image becomes fuzzy. As a result, while reducing picture noise, edge of the image must be protected. The wavelet algorithm technique is a time-frequency evaluation tool that adapts the frequency band selection based on the signal properties. The frequency band is then matched to the spectrum, improving time-frequency accuracy. The wavelet transform is designed in such a way that to remove noises in the signals, i.e., the low-frequency components have strong frequency resolution (Xizhi 2008).

8.2.3.6 Image Processing via Neural Network Algorithm

Multi-layered networks made up of neurons or nodes are known as neural networks. The neural network's fundamental processing units are these neurons. They are programmed to function similar to human brains. They gather information, teach themselves to detect patterns in it, and then forecast the outcome. The input layer, hidden layer, and output layer are the three layers of a simple neural network (Fig. 8.4).

The input layers read and receive the information, the output layer forecasts the outcome, and the hidden layers perform the majority of the computations.

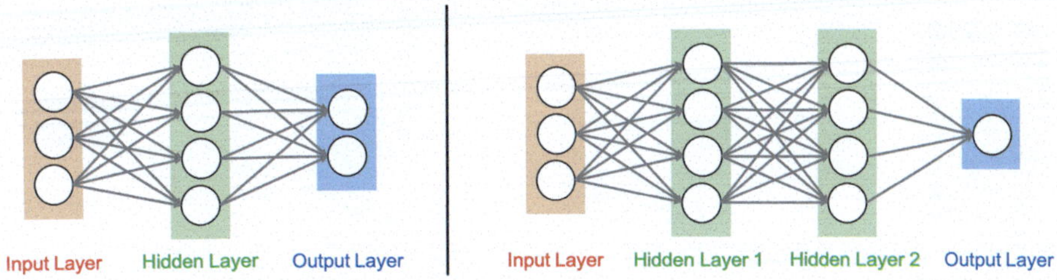

Fig. 8.4 A simple two (left) and three (right) layer neural network

Depending on the needs, the number of hidden layers could be changed. A neural network, on the other hand, should include at least one hidden layer.

8.2.3.7 Contrast Enhancement Algorithm for Color Images

The histogram equalization technique, which utilizes the histogram to increase picture contrast, and the adaptable histogram equalization algorithm, which adjusts to local changes, in contrast, are two types of color improvement algorithms. In addition to this, traditional image processing algorithms, namely, connected-component labeling algorithm, dithering, and half-toning algorithm, error diffusion algorithm, Floyd–Steinberg dithering algorithm, ordered dithering algorithm, Riemersma dithering algorithm, Elser difference-map algorithm, feature detection algorithm, the algorithm of Marr-Hildreth, the algorithm of Canny edge detector, the algorithm of Generalized Hough transform, the algorithm of blind deconvolution are also useful.

8.2.4 Neural Network Types

The mammalian brain's visual cortex is composed of neuronal cells that collect picture characteristics. Each neural cell collects distinct characteristics that aid picture comprehension. Animals' ability to comprehend images is an interesting experience for them, as well as a very straightforward job. However, there are many underlying complications in the technique for a machine to interpret a picture.

8.2.4.1 Convolutional Neural Network (CNN)

Animals sense the picture being captured by their eyes that is then analyzed by neurons and transmitted to the brain for understanding. CNN is a deep learning system that tries to mimic the optical circuitry of animals and is influenced by the cerebral cortex of the animal brain. CNNs offer a paradigm shift in the field picture comprehension, including picture categorization, segmentation, localization, recognition, and other tasks. The effectiveness of CNNs in picture comprehension is the principal cause for their widespread acceptance. CNN's are made up of convolutions with learnable hidden layers, comparable to the animal's neurons. The fundamental

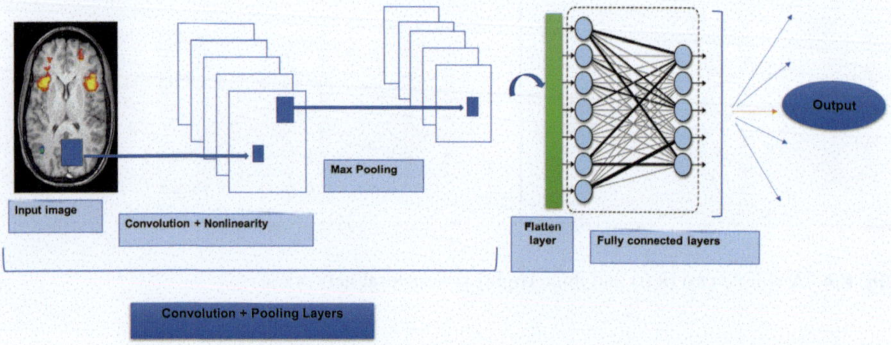

Fig. 8.5 A typical convolutional neural network (CNN) building for medical image processing

structural elements of the convolutional neural network are convolutional, pooling, and fully connected layers (Teuwen and Moriakov 2020), as shown in Fig. 8.5.

8.2.4.2 Convolutional (CONV) Layer

The central component of CNN, which is responsible for conducting convolutional layers operations. The goal of the CONV layer, which is modeled over the neural cells, is to collect characteristics including edges, color, texture, and gradient direction. Convolutional filters, often known as kernels, are learnable filters (matrix). When the whole picture is scanned, the kernel produces horizontal and vertical changes dependent on the stride rate (convolution process). The kernels are normalized over the width and height of the input volume through the forward run, and the dot product is computed between the filter entries and the input. Instinctively, the CNN builds filters that operate when it encounters edges and colors. Then the out of CONV layer is fed into the pooling layer.

8.2.4.3 Pooling Layer (POOL)

Pooling layer aids in reducing the amount of computing power required to process the data and is responsible for reducing the dimension. With the help of dimension reduction, it reduces the amount of processing power necessary to process the dataset. A nonlinear downward sampling of convolved features is performed by the pooling layer. It controls overfitting and overcomes translational and rotation variability and decreases dimensionality of pictures by accumulating information over space or component type. The splitting of its input into a collection of rectangular sections is the outcome of the pooling procedure. Depending on the type of pooling used, each patch is substituted with a single value. Pooling may be divided into two types: maximum pooling and average pooling. The largest value from the region covered by the kernel on the image is provided by maximum pooling. The average of most of the values in the region of the picture represented by the kernel is provided by average pooling.

8.2.4.4 Fully Connected (FC) Layer

FC layer works with a flattened input, which means that each input is linked to all the neurons. The classifier function is performed by the FC layer, which is accompanied by the sigmoid activation function. FC layer is comparable to an artificial neural network in that each node has connectivity to all of the inputs and that each connectivity has weights related to it. The output is the combination of all the inputs multiplied by the weights. FC layers are generally found at the end of CNN architectures and is mostly utilized to extract information from the images. In image classification, CNNs are commonly used to process each input image via a sequence of layers to obtain a probability value between 0 and 1.

8.2.5 Generative Adversarial Networks (GANs)

Generative adversarial networks (GANs) have a lot of interest in the field of computer vision owing to their capability to gather data information without explicitly modeling the probability distribution. The discriminator's adversarial deficit presents an effective approach to incorporate unlabeled data into training while still imposing a higher degree of uniformity. GAN is adapted for variety of situations, including domain adaptation, feature extraction, and image-to-image interpretation.

Generator and discriminator are the two models that make up GANs. The generator tries to create realistic-looking false pictures to deceive the discriminator. Discriminator develops the ability to differentiate between false and real pictures (it tries not to get fooled). Because the generator is not permitted to see actual pictures, it may generate bad results in the beginning, but the discriminator is allowed to see real images; moreover, they are mixed in with the generator's false images, which it must categorize as real or fake (Fig. 8.6). The generator is supplied some noise as input so that it can create various samples each time rather than the

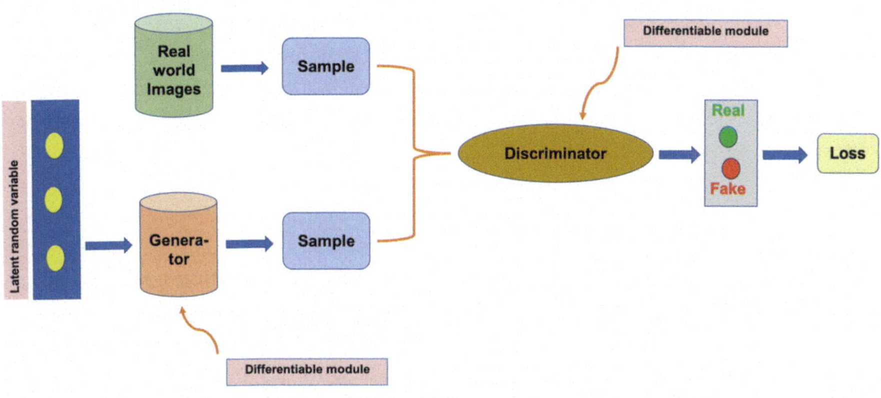

Fig. 8.6 Training of discriminator and generator

same kind of image. The generator attempts to enhance its results based on the discriminator's anticipated scores; after a given period, the generator would be able to create images that are more difficult to differentiate; at that point, the user will be pleased with the results. Discriminator also advances with each round of the generator, as it receives more and more realistic pictures. Deep convolutional GANs (DCGANs), conditional GANs (CGANs), StyleGANs, CycleGAN, DiscoGAN, GauGAN, etc. are some of the most common forms of GANs. These characteristics have caught the attention of biomedical imaging experts and have seen successful implementation in a wide range of conventional and innovative applications, including image restoration, segmentation, recognition, categorization, and cross-modality synthesis (Yi et al. 2019).

8.2.6 Medical Image Processing Tools

Medical imaging is the technique and art of creating image representations of the interior of a human body for clinical assessment and medical intervention using. The goal of medical imaging is to reveal hidden internal structures beneath the skin and bones, as well as to diagnose and treat sickness. Biomedical imaging also develops a database of normal anatomy and physiology, making it easier to spot abnormalities. Biomedical imaging of dissected organs and tissues is also feasible, although such procedures are usually classed as pathology instead of medical imaging. The most widely adopted medical image processing tools (software) are listed in Table 8.1.

8.2.6.1 Visualization Toolkit (VTK)

Visualization Toolkit is a free and open-source software platform for 3D computer graphics, image processing, and visualization. This tool provides VTK with professional assistance and business solutions and have a big impact on image processing (Pettit et al. 2012). VTK is also a cross-platform, permissively licensed toolkit for empirical data processing, visualization, and interpretation. Hanwell et al. (2015) claimed the use of VTK to evaluate the results of new and old graphics code on the same computers and cards. Scientific data may now be displayed in more detail than ever before because of significant advances in rendering rates and memory footprints.

8.2.6.2 Insight Toolkit (ITK)

ITK is an open-source, cross-platform system that provides cutting-edge segmentation and registration algorithms in two, three, and multiple dimensions. ITK specifies an "image iterator" to scan an image region during the region developing stage, which is an abstraction of a memory pointer, allowing for more efficient memory use. Liu et al. (2014) and Johnson and McCormick (2021) employed ITK suite for image processing applications.

Table 8.1 Medical image processing tools (software) and their significance

S. No.	Tools	Application	Reference
1	PIL (Python Image Library)/ pillow	It supports handling of wide range of image formats like PPM, JPEG, TIFF, GIF, PNG, and BMP	https://python-pillow.org/
2	NumPy	Flipping, extracting hidden features, and analyzing the images	https://numpy.org/install/
3	Visualization Toolkit	Suite of 3D interaction widgets, which supports parallel processing and integrates with various databases	https://vtk.org/
4	Insight Toolkit	Cutting-edge segmentation and registration algorithms in two, three, and multiple dimensions	https://itk.org/
5	FMRIB Software Library	Employed for eliminating additional cranial tissues from MR images	https://fsl.fmrib.ox.ac.uk/fsl/fslwiki
6	Statistical Parametric Mapping	Suite to analyze brain imaging data sequences including fMRI, PET, SPECT, EEG, and MEG	https://www.fil.ion.ucl.ac.uk/spm/
7	GIMIAS	Assists in the development of research and clinical software prototypes.	http://www.gimias.org/
8	NiftyReg	Registering efficient medical images.	http://cmictig.cs.ucl.ac.uk/wiki/index.php/NiftyReg
9	Elastix	Segmentation and Registration Toolkit	https://elastix.lumc.nl/
10	Advanced Normalization Tools	Cutting-edge picture registration, segmentation, and template creation capabilities	http://stnava.github.io/ANTs/
11	NiftySeg	EM-based picture Nifty segmentation and analysis	https://www.onworks.net/software/linux/app-nifty-seg
12	ITK-Snap	Manual segmentation and picture navigation, as well as semi-automatic delineation employing active contour techniques	http://www.itksnap.org/pmwiki/pmwiki.php
13	Medical Imaging Interaction Toolkit	Integrates the Insight Toolkit with the Visualization Toolkit	https://www.medfloss.org/node/267
14	NiftyRec	Provides coding for tomographic restoration	https://www.onworks.net/software/app-niftyrec?amp=0
15	NiftySim	To incorporate high-performance soft tissue simulation capabilities	https://www.onworks.net/software/windows/app-nifty-sim?amp=0
16	Camino	An open-source for diffusion MRI imaging and image processing	http://picsl.upenn.edu/software/camino/

(continued)

Table 8.1 (continued)

S. No.	Tools	Application	Reference
17	DTI-TK	Spatial normalization and map building that is tailored for morphometric analysis of white matter utilizing DTI data	http://dti-tk.sourceforge.net/pmwiki/pmwiki.php
18	QuantX	Assessment and characterization of breast abnormalities using magnetic resonance imaging	https://www.qlarityimaging.com/
19	ProFound AI	AI tool for the analysis of digital breast tomosynthesis (DBT) to detect malignant soft tissue densities and calcifications.	https://www.icadmed.com/profoundai.html
20	IDx-DR	AI diagnosis of diabetic retinopathy and macular edema	https://www.digitaldiagnostics.com/products/eye-disease/idx-dr/
21	OsteoDetect	Analyzes wrist radiographs (X-rays) using machine learning techniques to identify and highlight distal radius fractures	https://www.lify.io/3p-products/osteodetect
22	Computed tomography (CT)	CT is a procedure for computerized X-ray imaging, for instance, breast CT scanner.	https://www.nibib.nih.gov/science-education/science-topics/computed-tomography-ct
23	Triton (Gauss Surgical)	To estimate blood loss during surgery as collected in surgical sponges and dressings	https://www.gausssurgical.com/

8.2.6.3 FMRIB Software Library (FSL)

FSL was invented by Analysis Group, FMRIB, and OXFORD, UK (Smith et al. 2004) for analyzing functional MRI (fMRI), MRI, and DTI brain images. The Brain Extraction Tool (BET) from the FMRIb Software Library (FSL) eliminates additional cranial tissues from MR images Fagiolo demonstrated a recursive technique for obtaining brain pictures using BET (Fagiolo et al. 2008).

(a) **Statistical Parametric Mapping (SPM):** The design and evaluation of spatially extended statistical procedures to evaluate the hypotheses regarding functional imaging data are referred to as SPM. SPM software suite analyzes brain imaging data sequences including fMRI, PET, SPECT, EEG, and MEG. Penny et al. (Penny et al. 2007) have used SPM in the identification of brain abnormalities. Sowell et al. (Sowell et al. 2000) utilized SPM to look at statistical parametric maps of grey matter, white matter, and CSF differences between the groups. Premature schizophrenia patients exhibited larger ventricles, especially in the posterior horns of the lateral ventricles, as well as anomalies in the mid callosal, posterior cingulate, caudate, and thalamic regions.

(b) **Graphical Interface for Medical Image Analysis and Simulation (GIMIAS):** GIMIAS is a workflow-oriented framework for tackling sophisticated biomedical image processing and customized simulation issues that is expandable through the creation of problem-specific plug-ins (Larrabide et al. 2009). Qasrawi and Ivorra (2015) claimed the utility of GIMIAS in increasing cell membrane permeability with respect to high-intensity electric stimuli. In live cells, permeabilization enhancements could be used to improve the medication or DNA plasmid entry or to kill unwanted cells, and it's usually done by administering pulsed high voltage over needle electrodes.

(c) **NiftyReg:** NiftyReg was developed by Centre for Medical Image Computing, Department of Medical Physics, University College London (https://github.com/KCL-BMEIS/niftyreg/wiki). It provides linear and nonlinear registration and processing for two-dimensional (2D), three-dimensional (3D), and four-dimensional (3D) medical images. NiftyReg along with MIM programs was utilized in the measurement of gross lung tumor volume (Sugawara et al. 2017). The NiftyReg and MIM nonrigid registration algorithms were based on freeform distortion. Different deformable image registration (DIR)-based ventilation imaging was employed in the visualization of pulmonary ventilation function and validated using NiftyReg (Miyakawa et al. 2018).

(d) **Elastix:** A free software based on the popular Insight Segmentation and Registration Toolkit (ITK) to tackle medical picture registration issues. Elastix's modular structure enables users to rapidly create, evaluate, and analyze several registration techniques. The Elastix toolkit deformation image registration approach is implemented for volume of interest (VOI) and voxel-wise evaluation of longitudinal changes in fluorodeoxy glucose uptake in non-small cell lung carcinoma patients (Kerner et al. 2015).

(e) **Advanced Normalization Tools (ANTs):** Avants and Song developed ANT tool (Avants et al. 2011a) to extract data from large datasets and is extremely useful for organizing, analyzing, and displaying multidimensional data (Avants et al. 2011b). The National Institutes of Health (NIH) in the United States invests much in open-source data and tools to promote reproducibility in the biomedical image processing sciences. ANT is more suitable for quantitative morphometric analysis including cutting-edge picture registration, segmentation, and template creation capabilities. ANTs are employed to quantify the influence of cosine similarity on the linear and deformable components of template-based normalization research for the first time. This is the first project to employ a consistent transformation framework to assess the standalone influence of the similarity measure on optimum template creation and brain labeling in a repeatable manner.

(f) **NiftyNet:** A deep learning based medical image processing software package. It features image segmentation, data augmentation, image regression, image generation, and visualization options. Niftynet was employed for the evaluation of idiosyncrasies of medical images (2D and 3D) (Gibson et al. 2018).

(g) **ITK-Snap:** Yushkevich et al. (2006) created an open-source program called ITK-SNAP to bridge the gap between scientific advancements and clinical practice to make level set segmentation available to a wide variety of users, including those with little or no mathematical experience. Adaptive threshold segmentation and its robust implementation utilizing level set techniques are well-established conceptual methodologies that have been extensively researched in the image processing domain. ITK-Snap software application is useful segmenting entities in 3D medical pictures. It includes manual segmentation and picture navigation, as well as a semiautomatic delineation employing the active contour technique. Yushkevich's work explains the findings of validation trials conducted as part of an ongoing pediatric autism neuroimaging research.

(h) **Medical Imaging Interaction Toolkit (MITK):** Medical Imaging Interaction Toolkit integrates the Insight Toolkit (ITK) with the Visualization Toolkit (VTK), which is licensed under the BSD-Style license (Nolden et al. 2013). MITK makes it simple to combine algorithms produced by ITK with the visualizations made by VTK, and it expands these two toolkits with functionalities that are outside the scope of both. Lu et al. (2012) proposed an integrated system created by combining two free software projects, viz. IGSTK and MITK. The software integration reduced the time required to construct an image-guided surgical navigation system. Seitel et al. (2012) used MITK as a framework for supplying image processing techniques for an image-guided surgery navigation system of a medical image processor. The combination of sophisticated technologies enhanced the precision, safety, and durability of the image-guided surgical navigation system.

(i) **NiftyRec: NiftyRec** is a University College London software project that provides coding for tomographic restoration (Assaf and Alexander 2014). This program includes both global and a local registration algorithm using a block-matching method and used for the lung registration. The methodology utilized by the Niftyreggers crew, as well as the findings, is described by Modat et al. (2010).

(j) **NiftySim:** NiftySim is a freely accessible finite element toolbox developed by University College London which allows biomedical applications to incorporate high-performance nonlinear finite element solution for soft tissue simulation. NiftySim toolbox uses graphics processing unit (GPU) technology to deliver high-performance soft tissue simulation capabilities (Johnsen et al. 2015). In an easy-to-use package, the toolkit offers the possibility of running on a fast GPU hardware, as well as many constitutive models and solid element alternatives, membrane, and shell elements, and contact modeling tools. The toolkit is created through the total Lagrangian explicit dynamics (TLEDs) technique, which is being demonstrated to be fast and effective for modeling soft tissues.

(k) **Camino:** Camino is a software framework for diffusing MRI imaging that allows building processing pipelines of other program modules. The toolbox is maintained by the microstructure imaging group at UCL. Diffusion-MRI is a fast-developing research topic that has yielded a plethora of methods for analyzing white matter filament structure and brain diseases. Camino is a freeware for diffusion MRI researchers that makes a selection of this technology available and easy to use (Cook et al. 2006).

(l) **DTI-TK:** DTI-TK is a toolkit for spatial normalization and map building. DTI-TK is tailored for morphometric analysis of white matter utilizing diffusion tensor imaging (DTI) data (Keihaninejad et al. 2013). In modern days, DTI has been the preferred method for investigating white matter abnormalities in the developing brain.

8.3 Applications of Image Processing

Medical imaging focuses on exposing and discovering underlying structures that are buried internally under the skin and bones. It is also used to investigate, diagnose, recognize, and treat illness or disease. Medical imaging is useful for performing laparoscopic procedures to observe the inside portions of the body without having to open the body (non-invasive procedure). X-ray imaging uses CT scanners, ultrasound, and magnetic resonance imaging.

8.3.1 Digital Image Processing

The realm of technical endeavor has been revolutionized by digital image processing, and there is still much more study to be done in this area. In some manner, digital image processing has influenced practically every sector of the technical inclination. Table 8.2 depicts the primary applications of digital imaging.

8.3.1.1 Gamma-Ray Imaging

Gamma ray imaging, "functional imaging," shows how organs are working, not just their structure ("physiology" not just "anatomy"). It is widely used to learn about normal physiology (e.g., brain activity) in volunteers, or more commonly to diagnose specific illnesses. Nuclear medicine and astronomical observations are the

Table 8.2 The key usage of different digital rays

Gamma rays	X-rays	Ultraviolet band	Visible and infrared bands	Microwave band
1. Tumors 2. Heart defects 3. Kidney problems	1. Bones 2. Joints 3. Blood vessels 4. Teeth 5. Lungs 6. Mammography 7. Foreign objects	Forensic investigations	1. Hemorrhagic shock 2. Peripheral artery disease 3. Dental cavities 4. Kidney stone 5. Laryngeal abnormalities	Early-stage breast cancer identification

dominant uses of imaging based on these gamma rays (Ahad et al. 2018). This technique can be used for both diagnosis and prevention. Imaging with gamma rays has a wide range of functions, such as tumor imaging, infection imaging, bones imaging, thyroid imaging, cardiac functionality, brain imaging, and diagnosis of Alzheimer's.

Gamma ray imaging is widely adopted in radioisotope diagnostic laboratories, scientific research medical institutes, cancer and psychiatry departments, and other medical organizations (Gonzalez and Woods 2013). These scans are utilized to pinpoint the locations of infections in the bones (Fig. 8.7) (Wolfe 1997a).

8.3.1.2 X-Ray Imaging

X-ray technology is one of the oldest and most common forms of medical imaging. X-rays are a form of electromagnetic radiation that sends small waves through the body, which are then absorbed in different amounts depending on the materials they pass through. Materials like fat and muscle often appear as shades of gray. X-rays are dominantly used in medical diagnostics, industry, and astronomy (Fig. 8.8) (Sone et al. 1991; Yokoi et al. 1991).

8.3.1.3 Ultraviolet Band Imaging

It is a process of photographing and recording images using the ultraviolet (UV) spectrum. The UV light images are used for scientific, medical, or artistic purposes. These images help in viewing the structures which cannot be viewed under visible light. In recent years, there has been an increase in demand for high-quality imaging in the UV light area, particularly in the fields such as forensic investigations (Zhang et al. 2021) (Fig. 8.9).

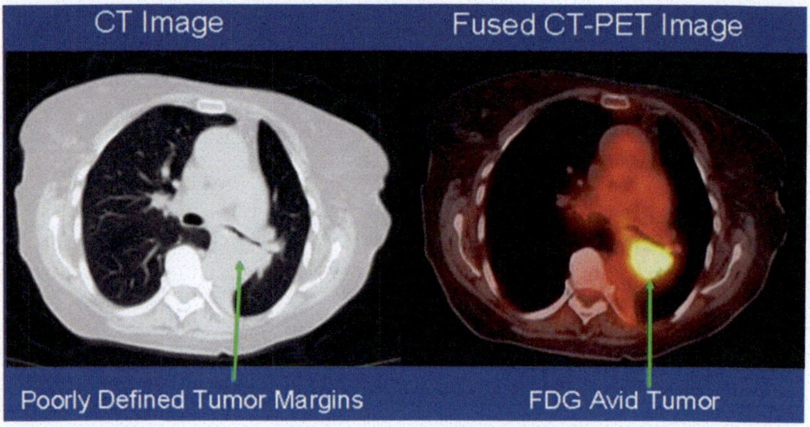

Fig. 8.7 An image by gamma rays (Open Source: National Institute of Biomedical Imaging and Bioengineering (NIBIB) website from the url: https://www.nibib.nih.gov/science-education/science-topics/nuclear-medicine)

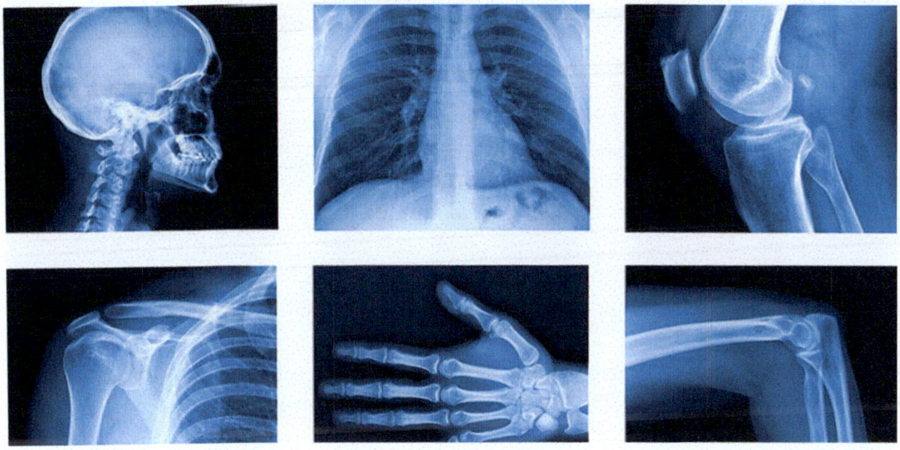

Fig. 8.8 An image by X-Ray (Open Source from website; https://whxray.com/services/x-ray/)

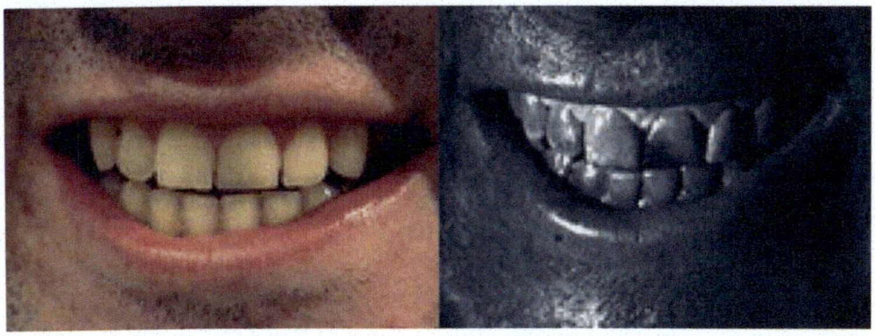

Fig. 8.9 A left incisor repaired with dental resin. ***Left:*** color, ***right:*** near-UV (Reprinted with permission from the *Photonics Handbook,* online at www.PhotonicsHandbook.com. Copyright 2021, Photonics Media, Pittsfield MA, https://www.photonics.com/a32169)

8.3.1.4 Visible and Infrared Bands Imaging

Remote sensing gave rise to hyperspectral imaging (HSI), also known as imaging spectrometer (Wolfe 1997b). HSI, as a new imaging modality for medical use, has a lot of promise for non-invasive illness diagnosis and surgical guidance. Multiple scattering and absorption of light given to biological tissue occur due to inhomogeneity of the biological structures and absorption principally non-hemoglobin, melanin, and other proteins (Zonios et al. 1999; Wang et al. 2008). The absorption, fluorescence, and scattering characteristics of tissues are thought to vary as the disease progresses. As a result, HSI captures quantitative diagnostic information on tissue pathology in the form of reflected, fluorescence, and transmitted light from the tissue (Fig. 8.10) (Lu and Fei 2014; Steelman et al. 2019; Ferris et al. 2001; Pierce et al. 2012). HSI can offer near-real-time images of biomarker information like oxyhemoglobin and deoxyhemoglobin and provide an assessment of tissue pathophysiology based on the spectral characteristics of different tissue (Panasyuk

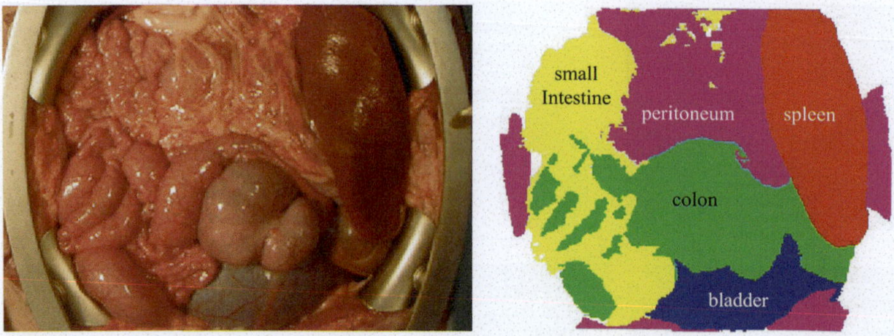

Fig. 8.10 Hyperspectral imaging (HSI) in capturing the tissue pathology of abdomen (Source: Creative Commons Attribution 3.0 unported license from SPIE digital library) (Lu and Fei 2014)

et al. 2007). As a result, HSI is being more widely used in medical diagnostics and image-guided surgery (Cancio et al. 2006; Zhang et al. 2020).

8.3.1.5 Microwave Band Imaging

Based on the huge differential in electrical characteristics between malignant tumor tissue and the surrounding normal breast tissue, ultrawideband (UWB) microwave imaging (Fig. 8.11) is a promising tool for early-stage breast cancer identification (Wang et al. 2014).

8.3.2 Biomedical Imaging System

X-ray, CT scan, magnetic resonance imaging (MRI), positron emission tomography (PET), and ultrasound pictures are used in this application. The system's principal uses include the detection of numerous ailments such as brain tumors, breast cancer, epilepsy, lung disorders, and heart diseases, among others. In the detection of brain cancers, the biomedical imaging system is commonly used. The brain is thought to be the control center of the nervous system. It is in charge of regulating all the functions of the human body. As a result, any anomaly in the brain will pose a threat to one's health (Lather and Singh 2021). The brain tumor is the result of unregulated and aberrant cell proliferation. It not only affects the brain's immediate cells, but it can also cause harm to other cells and tissues due to inflammation (McFaline-Figueroa and Lee 2018). Zhang et al. (Zhang et al. 2019) provided an automated technique that uses a hybrid approach of MRI, discrete wavelet transform (DWT), and K-means to detect and segment the brain tumors, allowing for accurate detection and treatment planning. GI endoscopy, which is used to examine the gastrointestinal tract and detect luminal pathology, is another application of medical imaging (Beckmann and Smith 2004).

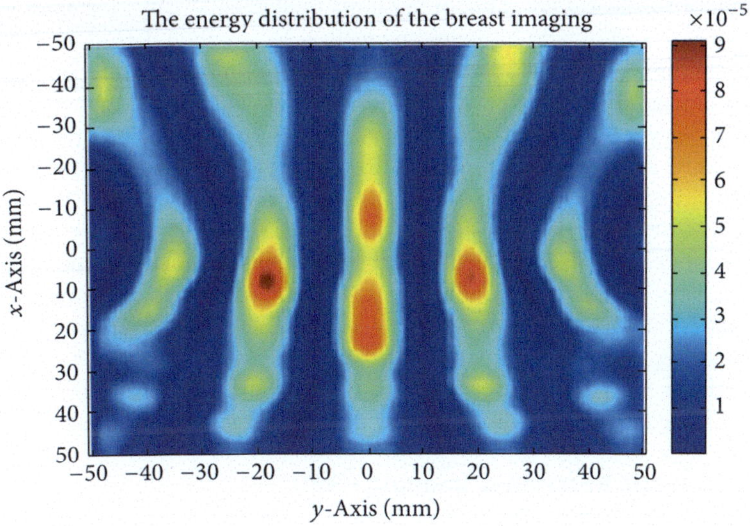

Fig. 8.11 An image of the microwave band (Source: Open access source from Hindawi) (Wang et al. 2014)

8.3.2.1 X-Ray

Low-intensity noisy gray-scale pictures with poor contrast and inadequate border representation are the common X-ray images (Tripathy and Swarnkar 2020). The feature extraction task is pretty difficult based on this low-quality image. Image enhancement techniques can be used to improve the efficiency of feature extraction and segmentation operations. For example, Hassanpour et al. (Hassanpour et al. 2015) proposed a new method for automatically enhancing the contrast of numerous high-resolution X-ray images. For the best imaging of hard tissues like bones and dentin, the morphological operators were utilized for contrast enhancement. The joint transforms of the top-hat and bottom-hat transform can be used to improve contrast. This notion is also investigated in the segmentation of retinal vessels. In this scenario, the combined approach can improve the image by determining the gradient magnitude value for autonomous structuring element size selection (Goulart et al. 2017). The efficiency of the procedure is determined by comparing it to the baselines in an experimental evaluation.

Furthermore, to improve the characterization of breast lesions, improved mammography (digital breast tomosynthesis) and X-ray imaging approach have been proposed (Helvie 2010). An X-ray tube projects a breast over a brief arc of movement in the breast tomosynthesis procedure (Liu et al. 2017). Understanding the X-ray scan characteristic is a vital challenge in developing accurate computer-aided diseases diagnoses (CADD). These algorithms are useful to decipher the images from CT scans, ECGs, ultrasounds, and MRIs.

Various image processing algorithms have enabled diverse edge detection techniques to develop a comprehensive solution for diagnosis analysis and recognition of medical scans. One of the most fundamental ways for extracting the boundary features of images of bones and teeth is to use edge detection algorithms. The strategies are carried out in practice by detecting contrast and sudden brightness changes in image pixels. An image enhancement method is proposed by Lyu et al. (2018) to overcome the medical image blurring problem and provide a more accurate and meaningful image. The approach is specifically designed to deblur the image using conditional generative adversarial networks (CGAN). The created convolutional neural network (CNN) is employed to recreate a sharp CT picture. The CNN model is trained to generate a sharp CT scan picture (output) from a blurred CT scan image (input), allowing it to produce a sharp image with correct feature extraction. Meanwhile, a critic network is trained in an adversarial way as the discriminator to assess the similarity between the created restored sharp CT scan picture and the sharp CT. Recent work by Kupyn et al., in 2018 introduces the concept of the generator and discriminator in recovering hazy CT images.

8.3.2.2 Magnetic Resonance Imaging (MRI)

The MRI scan is a highly effective and precise technology for scanning the entire body and detecting disease. In the majority of situations, it is utilized when other tests are no longer efficient in providing sufficient information for the proper diagnosis and therapy (Soni and Mahajan 2019; Hu et al. 2021; Scahill et al. 2003; Kak et al. 2002).

8.3.2.3 Ultrasound Imaging

High-frequency sound waves are used to create the ultrasound image of the body. In practice, sound waves are directed to the area of interest (region of interest), and the echoes returned are gathered to provide the physician with a "live" and real-time view of the area of interest. With reasonable clarity and precision, real-time images from an ultrasonic imaging system may show the movement and structure of the region, multiple viewpoints, intra-relationship and connectivity among surrounding components, blood flowing through blood arteries, and so on. The visuals are created using the statistical information obtained from the reflection of sound waves. The ultrasound images' inner image processing can offer a physician with more delicate information. As a result, ultrasound data is used to decipher the underlying unknown concerns that are critical (Contreras Ortiz et al. 2012; Hricak et al. 2011; Wei and Li 2014).

8.3.2.4 Computer Tomography (CT)

Computer tomography (CT) uses a series of digitally processed X-ray images obtained from different angles. CT approach creates a cross-sectional image of the target item, allowing the user to see interior areas without having to cut them out. It has become an important tool in medical imaging, supplementing X-rays and medical ultrasonography. CT scanning for the head is typically used to identify and detect infarction, tumors, calcifications, hemorrhages, and bone damage. This scan can also be used to detect the changes in the internals of the lungs, such as acute and chronic lung disease. Embolism is diagnosed using a pulmonary angiogram. One sensitive approach for diagnosing heart and abdominal disorders is the CT. In addition to providing a new dimension of patient-related data for radiation therapy, four-dimensional computed tomography also delivers a vast volume of data for additional processing and analysis (Abdalbari et al. 2013).

8.3.2.5 Endoscopy

Endoscopy's true value lies in its ability to "see within" for medical diagnosis. An instrument is used to inspect and check the inside portions of an empty organ or pit of the body, unlike other approaches. With the use of a cutting instrument attached to the end of the endoscope, a specialist may see into human bodies, and small treatments can be performed, if necessary. This procedure is known as keyhole surgery, and it leaves only a small scar on the human body. Every endoscope has two fiber bundles, one for illuminating the inner structure of the organ and the other for collecting light reflection. The endoscope is tubular-shaped optical equipment used to examine body problems that cannot be seen with the naked eye. Endoscopy is commonly utilized in pregnancy, plastic surgery, orthopedic surgery, and endoscopic spinal surgery, among other treatments. Endoscopes are also employed in other sectors, such as bomb disposal and the FBI, to perform surveillance in confined locations (Nam et al. 2012).

Stereo Endoscope

A stereo endoscope consists of two cameras installed on a single laparoscope. The images from these cameras are displayed on a video monitor. Stereoscopic technology aids in the improvement of surgical accuracy and operational requirements in the medical and clinical fields. With the support of realistic depth perception of 2D imaging technology, it reduces the operating time and ensures patient safety (Getty and Green 2007). Stereoscopic imaging is also used in other domains such as digital mammography, anatomy instruction, diabetic retinopathy, and minimally invasive surgical procedures (Gao et al. 2012).

8.3.2.6 Electrocardiography (ECG)

ECG uses electrodes on the chest and limbs of the body to record the electrical activity of the heart. The primary goal of ECG is to obtain information about the structure and function of the heart. Suspected pulmonary embolism, fainting or collapse, seizures, cardiac stress tests, and atrial flutter are all signals that electrocardiography should be performed. Critically ill patients, those undergoing general anesthesia, and patients with recurrent cardiovascular diseases all benefit from frequent ECG monitoring (Suk et al. 2014).

8.3.2.7 Positron Emission Tomography (PET)

PET is a tool for monitoring metabolic activities in the body. The technique uses a positron emission tracer embedded in a physiologically active molecule to detect emitted gamma rays indirectly. The 3D photos of tracer throughout the body are created using computer analysis.

8.4 Image Processing in Body Parts

8.4.1 Brain Imaging

Many studies reveal that categorization, anatomical segmentation of brain regions, and tumor identification in Alzheimer's disease (AD), a disease that affects people of all ages, have been undertaken to utilize AI in brain research. The Gaussian restricted Boltzmann machine (RBM) was used to find feature expressions in volume patches of MRI and PET images and correctly classified AD/mild cognitive impairment/ healthy controls (Payan and Montana 2015). In the categorization of Alzheimer's disease, the 3D CNN outperforms alternative algorithm classifiers (Hosseini-Asl et al. 2018; De Brébisson and Montana 2015). It uses CNN to automatically segment MR images of the human brain (Choi and Jin 2016). Deep CNN was used to segment the striatum, and the findings were compared to those obtained with the FreeSurfer (Chen et al. 2018).

Manual segmentation in the brain takes time and can result in individual differences, but automatic segmentation has substantial issues operating in complicated structures. The voxel-wise residual network (VoxResNet), which consists of twenty five deep layers, was designed and successfully segmented automatically (Nie et al. 2016). A 3D fully CNN was used and tested in a real pelvic CT/MRI data set to demonstrate the end-to-end nonlinear mapping from MR pictures to CT images (Li et al. 2014). Two volume CNNs improvised the input and output performance and the input and output forms showed magnificent performance in the MRI and PET images of AD Neuroimaging Initiative (ADNI) database (Yoon et al. 2019).

8.4.2 Chest Imaging

A de-CNN is built using the multiple-instance learning architecture to build a heat map of suspicious locations (Shin et al. 2016). Using CNN algorithms, a unique set of radiologic datasets of publicly available chest X-rays and reports were used to detect and report 17 unique patterns (Gao et al. 2016). The presence of interstitial patterns has been discovered using a segmentation-based label propagation method applied to a dataset of interstitial lung disease, and lung texture patterns have been identified using CNN (Gao et al. 2018). Rajkomar et al. (2017) described a method for classifying frontal and lateral chest X-ray pictures and automating metadata annotations using deep learning approaches. A new strategy for reducing false positives in automatic lung nodule detection in volumetric CT scans utilizing a 3D CNN has been proposed. Through a hierarchical architecture trained on 3D samples, 3D CNN can enter more spatial information and extract more representative features. The suggested algorithm received good competition metric scores, was thoroughly evaluated in the LUNA16 Challenge, and may be used to analyze the 3D PET data.

8.4.3 Breast Imaging

Because most mammograms are 2D and have a great amount of data, medical images can be successfully examined using deep learning. The detection and classification of tumor lesions, micro-calcifications, and risk-scoring tasks all contribute to the identification of breast cancer, which can be efficiently studied using CNN or RBM approaches. CNNs were employed for feature extraction and localization in the measurement of breast density (Fonseca et al. 2015) and a modified region proposal CNN (R-CNN) was used for localization. It has been claimed that U-net is utilized to segment breast and fibro-glandular tissue in MRI datasets, resulting in correct breast density calculations (Dalmış et al. 2017). A short-term risk assessment model was established with a predictive accuracy of 71.4% by computing the risk score using the mammographic X-ray picture as a risk prediction module multiple layer perception classifier (Qiu et al. 2016).

8.4.4 Cardiac Imaging

Left ventricular segmentation, slice categorization, imaging quality assessment, automated calcium scoring, coronary center line tracking, and super-resolution are all areas of cardiac artificial research (SR). Classification is mostly accomplished using 2D and 3D CNN techniques, while segmentation is accomplished using deep learning techniques such as the U-net segmentation algorithm. A novel image SR technique was used to reconstruct the high-resolution 3D volume in the 2D image stack (Avendi et al. 2016). The image quality is better than the SR approach due to the CNN model's computational efficiency, but SR-CNN is better for image segmentation and motion tracking (Qiu et al. 2020). When the ROI is evaluated as a coronary artery calcification candidate exceeding 130 HU, it has been shown that deep learning can identify the low-dose chest CT with high accuracy using multi-stream CNN (3 views) (Lessmann et al. 2016). 3D CNN and multi-stream 2D CNN were used to identify the coronary calcium in gated cardiac CT angiography (Wolterink et al. 2016).

8.4.5 Musculoskeletal Imaging

Deep learning algorithms scan musculoskeletal pictures for segmentation and diagnosis of bone, joint, and soft tissue disorders. Several deep learning systems for autonomously estimating the age of skeletal bones using X-ray imaging have been developed, and their accuracy has been demonstrated by an average disparity of roughly 0.8 years (Spampinato et al. 2017). A 3D CNN architecture was created to execute supervised segmentation of vertebral bodies from 3D MR spine images automatically, with a "Dice" similarity coefficient of 93.4%. Due to the variety of spine shapes and postures, automatic spine recognition, which includes spine location identification and multiple picture naming, necessitates a vast amount of image data and is challenging to recognize. The posture of the spine was automatically rectified to process the image using a deep learning architecture termed modified deep convolution network (Cai et al. 2016; Miao et al. 2016).

8.5 Image-Based Profiling in Drug Discovery

There are two techniques to an experimental design that are fundamentally different in terms of probes and image processing when it comes to the image-based phenotyping of perturbations (Neumann et al. 2006).

8.5.1 Phenotype Screening

The first way entails employing screening programs (sometimes referred to as high-content or phenotypic screening) to identify drugs or treatment targets that affect pre-defined, specific traits. Image-based phenotypic screening, for example, has been successful in discovering targets and medicines that change characteristics like specie subcellular localization (Fig. 8.12) (Schirle and Jenkins 2016).

8.5.2 High-Throughput Imaging

The second application of high-throughput imaging in drug development is the more widespread characterization of perturbations. Multiplexed fluorescent dyes and fluorescently tagged antibodies are used to stain the cells and subcellular structures to view or "paint" them. Automated picture acquisition and analysis are applied in the phenotype profiling (Schenone et al. 2013). Computer vision can extract multivariate feature vectors of cell morphology such as cell size, shape, texture, and staining intensity without further human input (Fig. 8.13). All large-scale experiments described so far used segmentation techniques to correctly characterize cellular outlines before feature extraction. Single-cell or cell population profiles are then assessed to find correlations in data sets with tens of thousands of perturbations. Using high-throughput imaging, small molecule profiling in human cancer cells has been successfully used to group the small molecules with similar activity and mode of action (Perlman et al. 2004).

8.5.3 Clustering

The concept of clustering tiny compounds based on phenotypic profile similarity was broadened to include a correlation between phenotypic responses and chemical structural similarity (Young et al. 2008). A variety of drug research applications have successfully used the basic method of grouping tiny molecules based on phenotypic profile similarities. By linking the action of tiny molecules with the presence or

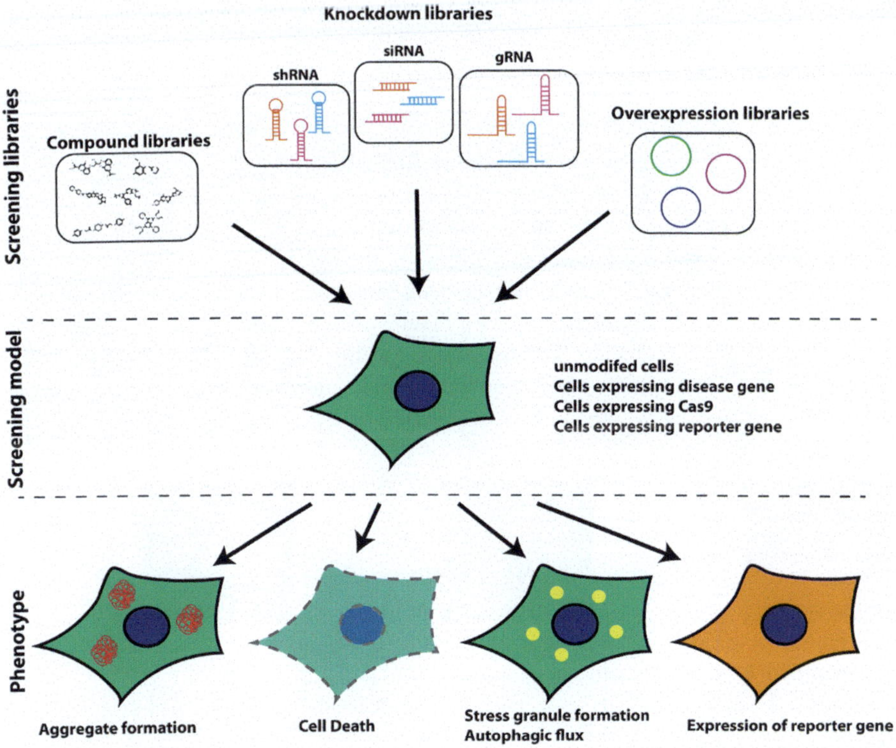

Fig. 8.12 An image of cell morphological changes. (Reprinted (adapted) with permission from Journal of Medicinal Chemistry (Brown and Wobst 2020). Copyright American Chemical Society)

absence of loss-of-function genes, chemical genetics approaches enable a systematic drug repurposing survey (Breinig et al. 2015). Off-targets and drug-target combinations with repurposing potential could be identified using annotated libraries of pharmacologically active small molecules. Small molecules can be unbiasedly screened for candidates that revert a disease-associated cellular phenotype to the wild-type phenotype, and small molecules can be linked to a new target by comparing their induced phenotype to that of a reference molecule.

8.5.4 MOA Profiling

MOA profiling of natural product libraries and enrichment of huge chemical libraries for effective and diverse subsets are two further applications. Studies have established image-based profiling for toxicity profiling, using model systems such as 3D liver cell culture or induced stem cell-derived cardiomyocytes (Tham et al. 2019).

Another application is the unbiased finding of hazardous small molecules in vast natural product libraries. After further refinement of image-based profiling assays

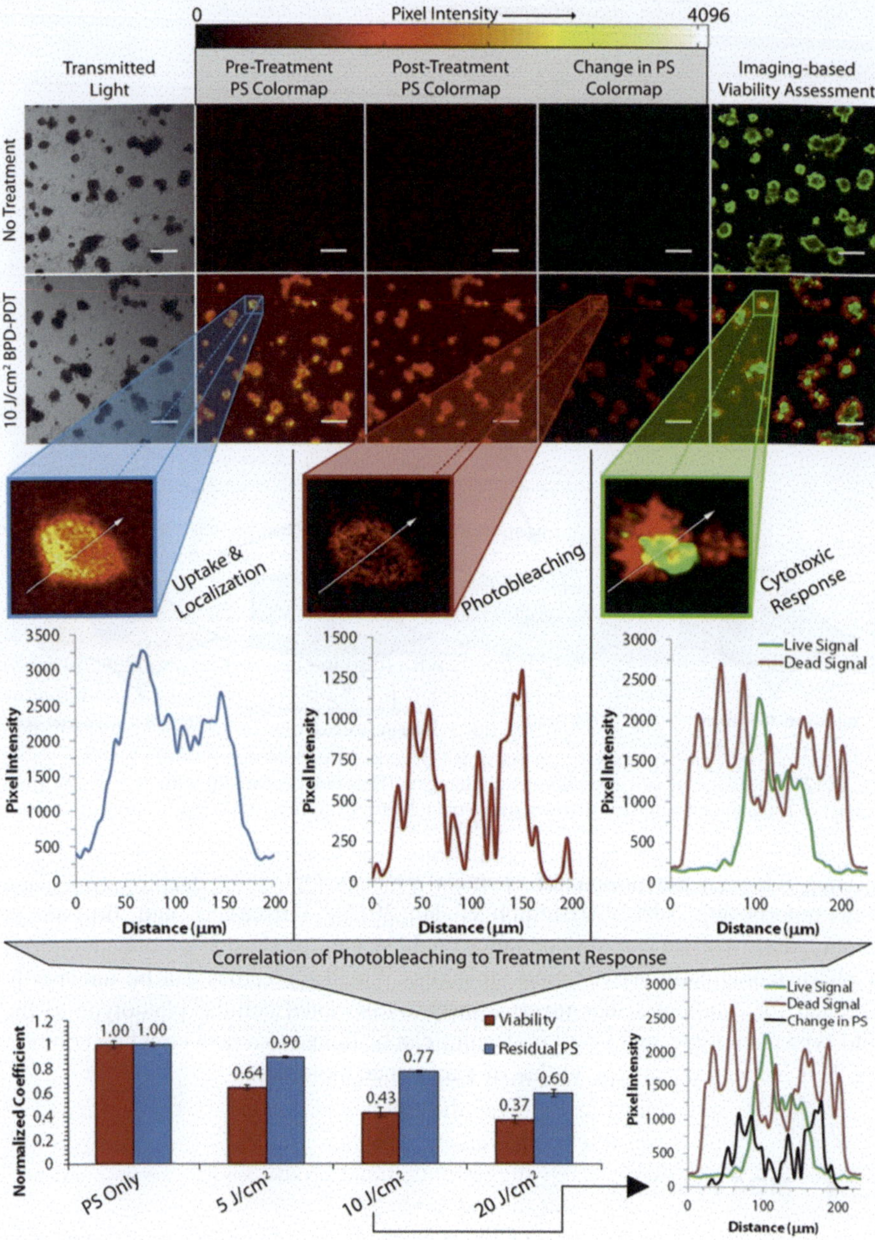

Fig. 8.13 A high-throughput imaging (Source: Open access from Frontiers in Physics (Evans 2015))

and analysis approaches, the technology became more accessible to more complex drug discovery investigations, as evidenced by a recent study in which 50,000 compounds were profiled in pluripotent stem cells to complement a target of other studies.

Image-based phenotypic profiling was recently used to characterize small medication function in patient-derived primary cells or 3D cell culture models. Booij et al. (2017) used image-based profiling to analyze the phenotypic effects of kinase inhibitors in a 3D cell culture model of polycystic kidney disease. The image-based profiling is employed to identify tiny drugs that could overcome resistance to the BCL2 inhibitor Venetoclax in primary CLL cells (Oppermann et al. 2016).

8.5.5 Image-Based Profiling and Chemical Genetics

Chemical genetics techniques have been pioneered in model organisms to examine the drug mechanism of action as well as gene activity. Libraries of loss-of-function or gain-of-function mutant strains were screened against chemical libraries to measure the genotype-dependent changes in fitness. Comparing phenotypic responses of mutant strains across chemical perturbations revealed the potential to pinpoint genes with similar functions as well as the bioactivity and cellular targets of tiny molecules (Cacace et al. 2017). Complex experimental and analysis frameworks have been established for chemical genetic investigations in model organisms such as yeast or bacteria. A recent study in *Saccharomyces cerevisiae* used an algorithm to generate a list of query mutant strains from genetic interaction data.

The goal of screening a subset of deletion strains against various chemical libraries was to functionally annotate tiny molecules across relevant biological processes without having to go through the complete deletion strain library. This concept has been used in pharmacogenomics studies in human cancer cell lines by combining the drug response and cell line genetic background with pharmacogenomics investigations in human cancer cell lines (Parsons et al. 2006). Several large-scale pharmacogenomics studies used univariate phenotypic readouts based on the cell viability assays that could be applied to a large panel of cell lines, although not all univariate cell variability assessments are equal (Gilbert et al. 2011).

Image-based chemical genetics has a lot of promise for identifying and investigating small molecules in human cell-based systems. Image-based profiling was used to systematically discover chemical genetic interactions in a smaller panel of human cancer cell lines (Caie et al. 2010; Warchal et al. 2016).

8.6 Conclusion

Various aspects can influence the success, efficiency, and quality of a project, but selecting the appropriate tools is one of the most essential—possibly save time and resources while achieving the greatest outcomes. The application of artificial intelligence techniques for image processing can help achieve the best results in health care. In particular, deep learning is one of the technologies which is altering the globe with its vast terminology and breakthroughs in the field of image processing. Different image processing procedures may be performed utilizing morphological, Gaussian, Fourier, and Wavelet algorithms which are applicable with image processing packages including OpenCV, Mahotas, and sci-kit-learn. In particular, neural networks for computer vision, such as CNN and GANs, are widely applicable for processing medical images. The existing and upcoming medical image processing tools find their application in supporting the physicians in the diagnosis and prognosis of the disease. Each technology for medical image processing has its own set of restrictions. Pharmacogenomics and phenotype studies combined with AI-supported image processing tools can be a promising method for the diagnosis and prognosis of illness more effectively and efficiently escalating the health care system to the very next level.

References

Abdalbari A, Huang X, Ren J (2013) Endoscopy-MR image fusion for image guided procedures. Int J Biomed Imaging 2013:472971. https://doi.org/10.1155/2013/472971

Abdallah YMY (2016) Increasing of edges recognition in cardiac scintography for ischemic patients. J Biomed Eng Med Imaging 2(6):39. https://doi.org/10.14738/jbemi.26.1697

Abdallah YMY (2018) Improvement of sonographic appearance using HAT-TOP methods. J Biomed Eng Med Imaging 5(5):13. https://doi.org/10.14738/jbemi.55.5283

Ahad MAR, Kobashi S, Tavares JMRS (2018) Advancements of image processing and vision in healthcare. J Healthcare Eng 2018:8458024. https://doi.org/10.1155/2018/8458024

Asokan A, Anitha J, Ciobanu M, Gabor A, Naaji A, Hemanth DJ (2020) Image processing techniques for analysis of satellite images for historical maps classification—an overview. Appl Sci 10(12):4207. https://doi.org/10.3390/app10124207

Assaf Y, Alexander DC (2014) Chapter 3.3—advanced methods to study white matter microstructure. In: Cohen-Adad J, C. A. M. B. T.-Q. M. R. I. of the S. C. Wheeler-Kingshott (Eds.) Academic, pp 156–163. https://doi.org/10.1016/B978-0-12-396973-6.00011-3

Avants BB, Tustison N, Song G (2011a) Advanced normalization tools (ANTS). www.picsl.upenn.edu/ANTS

Avants BB, Tustison NJ, Song G, Cook PA, Klein A, Gee JC (2011b) A reproducible evaluation of ANTs similarity metric performance in brain image registration. Neuroimage 54(3):2033–2044. https://doi.org/10.1016/j.neuroimage.2010.09.025

Avendi MR, Kheradvar A, Jafarkhani H (2016) A combined deep-learning and deformable-model approach to fully automatic segmentation of the left ventricle in cardiac MRI. Med Image Anal 30:108–119. https://doi.org/10.1016/j.media.2016.01.005

Balter S, Hopewell JW, Miller DL, Wagner LK, Zelefsky MJ (2010) Fluoroscopically guided interventional procedures: a review of radiation effects on patients' skin and hair. Radiology 254(2):326–341. https://doi.org/10.1148/radiol.2542082312

Barisoni L, Lafata KJ, Hewitt SM, Madabhushi A, Balis UGJ (2020) Digital pathology and computational image analysis in nephropathology. Nat Rev Nephrol 16(11):669–685. https://doi.org/10.1038/s41581-020-0321-6

Beckmann CF, Smith SM (2004) Probabilistic independent component analysis for functional magnetic resonance imaging. IEEE Trans Med Imaging 23(2):137–152. https://doi.org/10.1109/TMI.2003.822821

Belle A, Thiagarajan R, Soroushmehr SMR, Navidi F, Beard DA, Najarian K (2015) Big data analytics in healthcare. Biomed Res Int 2015:370194. https://doi.org/10.1155/2015/370194

Benjamens S, Dhunnoo P, Meskó B (2020) The state of artificial intelligence-based FDA-approved medical devices and algorithms: an online database. NPJ Digit Med 3(1):118. https://doi.org/10.1038/s41746-020-00324-0

Bohr A, Memarzadeh K (2020) The rise of artificial intelligence in healthcare applications. Artif Intellig Healthcare, 25–60. https://doi.org/10.1016/B978-0-12-818438-7.00002-2

Booij TH, Bange H, Leonhard WN, Yan K, Fokkelman M, Kunnen SJ, Dauwerse JG, Qin Y, van de Water B, van Westen GJP, Peters DJM, Price LS (2017) High-Throughput Phenotypic Screening of Kinase Inhibitors to Identify Drug Targets for Polycystic Kidney Disease. SLAS Discov 22(8):974–984. https://doi.org/10.1177/2472555217716056. Epub 2017 Jun 23. PMID: 28644734; PMCID: PMC5574491

Breinig M, Klein FA, Huber W, Boutros M (2015) A chemical-genetic interaction map of small molecules using high-throughput imaging in cancer cells. Mol Syst Biol 11(12):846. https://doi.org/10.15252/msb.20156400

Brown DG, Wobst HJ (2020) Opportunities and challenges in phenotypic screening for neurodegenerative disease research. J Med Chem 63(5):1823–1840. https://doi.org/10.1021/acs.jmedchem.9b00797

Cacace E, Kritikos G, Typas A (2017) Chemical genetics in drug discovery. Curr Opin Syst Biol 4: 35–42. https://doi.org/10.1016/j.coisb.2017.05.020

Cai Y, Landis M, Laidley DT, Kornecki A, Lum A, Li S (2016) Multi-modal vertebrae recognition using transformed deep convolution network. Comput Med Imaging Graph 51:11–19. https://doi.org/10.1016/j.compmedimag.2016.02.002

Caie PD, Walls RE, Ingleston-Orme A, Daya S, Houslay T, Eagle R, Roberts ME, Carragher NO (2010) High-content phenotypic profiling of drug response signatures across distinct cancer cells. Mol Cancer Therap 9(6):1913–1926. https://doi.org/10.1158/1535-7163.MCT-09-1148

Cancio LC, Batchinsky AI, Mansfield JR, Panasyuk S, Hetz K, Martini D, Jordan BS, Tracey B, Freeman JE (2006) Hyperspectral imaging: a new approach to the diagnosis of hemorrhagic shock. J Trauma 60(5):1087–1095. https://doi.org/10.1097/01.ta.0000217357.10617.3d

Carovac A, Smajlovic F, Junuzovic D (2011) Application of ultrasound in medicine. Acta Inform Med 19(3):168–171. https://doi.org/10.5455/aim.2011.19.168-171

Chen H, Dou Q, Yu L, Qin J, Heng PA (2018) VoxResNet: deep voxelwise residual networks for brain segmentation from 3D MR images. Neuroimage 170:446–455. https://doi.org/10.1016/j.neuroimage.2017.04.041

Choi H, Jin KH (2016) Fast and robust segmentation of the striatum using deep convolutional neural networks. J Neurosci Methods 274:146–153. https://doi.org/10.1016/j.jneumeth.2016.10.007

Computed tomography (CT), US-Food and Drug Administration (US-FDA). https://www.fda.gov/radiation-emitting-products/medical-x-ray-imaging/computed-tomography-ct. Accessed 15 July 2021

Contreras Ortiz SH, Chiu T, Fox MD (2012) Ultrasound image enhancement: a review. Biomed Signal Process Control 7(5):419–428. https://doi.org/10.1016/j.bspc.2012.02.002

Cook PA, Bai Y, Nedjati-Gilani SK, Seunarine KK, Hall MG, Parker GJ, Alexander DC (2006) Camino: open-source diffusion-MRI reconstruction and processing. In: 14th scientific meeting of the international society for magnetic resonance in medicine, vol. 2759. Seattle

Currie G, Hawk KE, Rohren E, Vial A, Klein R (2019) Machine learning and deep learning in medical imaging: intelligent imaging. J Med Imaging Radiat Sci 50(4):477–487. https://doi.org/10.1016/j.jmir.2019.09.005

Dalmış MU, Litjens G, Holland K, Setio A, Mann R, Karssemeijer N, Gubern-Mérida A (2017) Using deep learning to segment breast and fibroglandular tissue in MRI volumes. Med Phys 44(2):533–546. https://doi.org/10.1002/mp.12079

Dang X, Bardhan NM, Qi J, Gu L, Eze NA, Lin CW, Kataria S, Hammond PT, Belcher AM (2019) Deep-tissue optical imaging of near cellular-sized features. Sci Rep 9(1):3873. https://doi.org/10.1038/s41598-019-39502-w

Dash S, Shakyawar SK, Sharma M, Kaushik S (2019) Big data in healthcare: management, analysis and future prospects. J Big Data 6(1):54. https://doi.org/10.1186/s40537-019-0217-0

Davenport T, Kalakota R (2019) The potential for artificial intelligence in healthcare. Future Healthcare J 6(2):94–98. https://doi.org/10.7861/futurehosp.6-2-94

De Brébisson A, Montana G (2015) Deep neural networks for anatomical brain segmentation. In: IEEE computer society conference on computer vision and pattern recognition workshops, pp 20–28. https://doi.org/10.1109/CVPRW.2015.7301312

Deserno Né Lehmann TM, Handels H, Maier-Hein Né Fritzsche KH, Mersmann S, Palm C, Tolxdorff T, Wagenknecht G, Wittenberg T (2013) Viewpoints on medical image processing: from science to application. Curr Med Imaging Rev 9(2):79–88. https://doi.org/10.2174/1573405611309020002

DiCarlo JJ, Zoccolan D, Rust NC (2012) How does the brain solve visual object recognition? Neuron 73(3):415–434. https://doi.org/10.1016/j.neuron.2012.01.010

Donya M, Radford M, ElGuindy A, Firmin D, Yacoub MH (2014) Radiation in medicine: origins, risks and aspirations. Global Cardiol Sci Prac 2014(4):437–448. https://doi.org/10.5339/gcsp.2014.57

Duggan MA, Anderson WF, Altekruse S, Penberthy L, Sherman ME (2016) The surveillance, epidemiology, and end results (SEER) program and pathology: toward strengthening the critical relationship. Am J Surg Pathol 40(12):e94–e102. https://doi.org/10.1097/PAS.0000000000000749

Evans CL (2015) Three-dimensional in vitro cancer spheroid models for photodynamic therapy: strengths and opportunities. Front Phys 3:15. https://www.frontiersin.org/article/10.3389/fphy.2015.00015

Fagiolo G, Waldman A, Hajnal JV (2008) A simple procedure to improve FMRIb software library brain extraction tool performance. Br J Radiol 81(963):250–251. https://doi.org/10.1259/bjr/12956156

Ferris DG, Lawhead RA, Dickman ED, Holtzapple N, Miller JA, Grogan S, Bambot S, Agrawal A, Faupel ML (2001) Multimodal hyperspectral imaging for the noninvasive diagnosis of cervical neoplasia. J Low Genit Tract Dis 5(2):65–72. https://doi.org/10.1046/j.1526-0976.2001.005002065.x

Fonseca P, Mendoza J, Wainer J, Ferrer J, Pinto J, Guerrero J, Castaneda B (2015) Automatic breast density classification using a convolutional neural network architecture search procedure. In: Proc SPIE, pp 9414. https://doi.org/10.1117/12.2081576

Fukuchi T, Okauchi T, Shigeta M, Yamamoto S, Watanabe Y, Enomoto S (2017) Positron emission tomography with additional γ-ray detectors for multiple-tracer imaging. Med Phys 44(6):2257–2266. https://doi.org/10.1002/mp.12149

Gao Z, Wu J, Zhou J, Jiang W, Feng L (2012) Design of ECG signal acquisition and processing system. In: 2012 international conference on biomedical engineering and biotechnology, pp 762–764. https://doi.org/10.1109/iCBEB.2012.128

Gao M, Xu Z, Lu L, Wu A, Nogues I, Summers RM, Mollura DJ (2016) Segmentation label propagation using deep convolutional neural networks and dense conditional random field. In: 2016 IEEE 13th international symposium on biomedical imaging, pp 1265–1268. https://doi.org/10.1109/ISBI.2016.7493497

Gao M, Bagci U, Lu L, Wu A, Buty M, Shin H-C, Roth H, Papadakis GZ, Depeursinge A, Summers RM, Xu Z, Mollura DJ (2018) Holistic classification of CT attenuation patterns for interstitial lung diseases via deep convolutional neural networks. Comput Methods Biomech Biomed Eng Imaging Vis 6(1):1–6. https://doi.org/10.1080/21681163.2015.1124249

Getty DJ, Green PJ (2007) Clinical applications for stereoscopic 3-D displays. J Soc Inf Disp 15:377–384. https://doi.org/10.1889/1.2749323

Gibson E, Li W, Sudre C, Fidon L, Shakir DI, Wang G et al (2018) NiftyNet: a deep-learning platform for medical imaging. Comput Methods Prog Biomed 158:113–122. https://doi.org/10.1016/j.cmpb.2018.01.025

Gilbert DF, Erdmann G, Zhang X, Fritzsche A, Demir K, Jaedicke A, Muehlenberg K, Wanker EE, Boutros M (2011) A novel multiplex cell viability assay for high-throughput RNAi screening. PLoS One 6(12):e28338. https://doi.org/10.1371/journal.pone.0028338

Gonzalez RC, Woods RE (2013) Digital image processing, 3rd edn. Pearson International Edition, Upper Saddle River

Goulart JT, Bassani RA, Bassani JWM (2017) Application based on the Canny edge detection algorithm for recording contractions of isolated cardiac myocytes. Comput Biol Med 81:106–110. https://doi.org/10.1016/j.compbiomed.2016.12.014

Hanwell MD, Martin KM, Chaudhary A, Avila LS (2015) The visualization toolkit (VTK): rewriting the rendering code for modern graphics cards. SoftwareX 1–2:9–12. https://doi.org/10.1016/j.softx.2015.04.001

Hassanpour H, Samadiani N, Mahdi Salehi SM (2015) Using morphological transforms to enhance the contrast of medical images. Egypt J Radiol Nucl Med 46(2):481–489. https://doi.org/10.1016/j.ejrnm.2015.01.004

Helvie MA (2010) Digital mammography imaging: breast tomosynthesis and advanced applications. Radiol Clin North Am 48(5):917–929. https://doi.org/10.1016/j.rcl.2010.06.009

Hosny A, Parmar C, Quackenbush J, Schwartz LH, Aerts HJWL (2018) Artificial intelligence in radiology. Nat Rev Cancer 18(8):500–510. https://doi.org/10.1038/s41568-018-0016-5

Hosseini-Asl E, Ghazal M, Mahmoud A, Aslantas A, Shalaby AM, Casanova MF, El-Baz A (2018) Alzheimer's disease diagnostics by a 3D deeply supervised adaptable convolutional network. Front Biosci 23:584–596. https://doi.org/10.2741/4606

Hricak H, Brenner DJ, Adelstein SJ, Frush DP, Hall EJ, Howell RW, McCollough CH, Mettler FA, Pearce MS, Suleiman OH, Thrall JH, Wagner LK (2011) Managing radiation use in medical imaging: a multifaceted challenge. Radiology 258(3):889–905. https://doi.org/10.1148/radiol.10101157

Hsu W, Markey MK, Wang MD (2013) Biomedical imaging informatics in the era of precision medicine: progress, challenges, and opportunities. J Am Med Inform Assoc 20(6):1010–1013. https://doi.org/10.1136/amiajnl-2013-002315

Hu J, Qing Z, Liu R, Zhang X, Lv P, Wang M, Wang Y, He K, Gao Y, Zhang B (2021) Deep learning-based classification and voxel-based visualization of frontotemporal dementia and Alzheimer's disease. Front Neurosci 14:1468. https://www.frontiersin.org/article/10.3389/fnins.2020.626154

Iranmakani S, Mortezazadeh T, Sajadian F, Ghaziani MF, Ghafari A, Khezerloo D, Musa AE (2020) A review of various modalities in breast imaging: technical aspects and clinical outcomes. Egypt J Radiol Nucl Med 51(1):57. https://doi.org/10.1186/s43055-020-00175-5

Jiang F, Jiang Y, Zhi H, Dong Y, Li H, Ma S, Wang Y, Dong Q, Shen H, Wang Y (2017) Artificial intelligence in healthcare: past, present and future. Stroke Vasc Neurol 2(4):230–243. https://doi.org/10.1136/svn-2017-000101

Johnsen SF, Taylor ZA, Clarkson MJ, Hipwell J, Modat M, Eiben B, Han L, Hu Y, Mertzanidou T, Hawkes DJ, Ourselin S (2015) NiftySim: a GPU-based nonlinear finite element package for simulation of soft tissue biomechanics. Int J Comput Assist Radiol Surg 10(7):1077–1095. https://doi.org/10.1007/s11548-014-1118-5

Johnson HJ, McCormick MM (2021) The ITK software guide book 1: introduction and development guidelines fourth edition updated for ITK version 5.2.0. https://itk.orghttps//discourse.itk.org/

Kak AC, Slaney M, Wang G (2002) Principles of computerized tomographic imaging. Med Phys 29(1):107–107. https://doi.org/10.1118/1.1455742

Keihaninejad S, Zhang H, Ryan NS, Malone IB, Modat M, Cardoso MJ, Cash DM, Fox NC, Ourselin S (2013) An unbiased longitudinal analysis framework for tracking white matter changes using diffusion tensor imaging with application to Alzheimer's disease. Neuroimage 72:153–163. https://doi.org/10.1016/j.neuroimage.2013.01.044

Kerner G, Fischer A, Koole M, Pruim J, Groen H (2015) Evaluation of elastix-based propagated align algorithm for VOI- and voxel-based analysis of longitudinal 18F-FDG PET/CT data from patients with non-small cell lung cancer (NSCLC). EJNMMI Res 5:15. https://doi.org/10.1186/s13550-015-0089-z

Kupyn O, Budzan V, Mykhailych M, Mishkin D, Matas J (2018) DeblurGAN: blind motion deblurring using conditional adversarial networks, 8183–8192. https://doi.org/10.1109/CVPR.2018.00854

Larrabide I et al (2009) GIMIAS: an open source framework for efficient development of research tools and clinical prototypes. In: Ayache N, Delingette H, Sermesant M (Eds.) Functional imaging and modeling of the Heart. FIMH 2009. Lecture notes in computer science, vol. 5528. Springer, Berlin, Heidelberg https://doi.org/10.1007/978-3-642-01932-6_45

Lather M, Singh P (2021) Image processing: what, how and future. In: Singh V, Asari V, Kumar S, Patel R (eds) Computational methods and data engineering. Advances in intelligent systems and computing, vol 1227. Springer, Singapore. https://doi.org/10.1007/978-981-15-6876-3_23

Latif A, Rasheed A, Sajid U, Ahmed J, Ali N, Ratyal NI, Zafar B, Dar SH, Sajid M, Khalil T (2019) Content-based image retrieval and feature extraction: a comprehensive review. Math Probl Eng 2019:9658350. https://doi.org/10.1155/2019/9658350

Lessmann N, Išgum I, Setio AAA, de Vos BD, Ciompi F, de Jong PA, Oudkerk M, Mali WPTM, Viergever MA, van Ginneken B (2016) Deep convolutional neural networks for automatic coronary calcium scoring in a screening study with low-dose chest CT. In: Proc SPIE, 9785. https://doi.org/10.1117/12.2216978

Li R, Zhang W, Suk HI, Wang L, Li J, Shen D, Ji S (2014) Deep learning based imaging data completion for improved brain disease diagnosis. Med Image Comput Comput Assist Interv 17(3):305–312. https://doi.org/10.1007/978-3-319-10443-0_39

Liu Y, Kot A, Drakopoulos F, Yao C, Fedorov A, Enquobahrie A, Clatz O, Chrisochoides N (2014) An ITK implementation of a physics-based non-rigid registration method for brain deformation in image-guided neurosurgery. Front Neuroinform 8:33. https://doi.org/10.3389/fninf.2014.00033

Liu F, Hernandez-Cabronero M, Sanchez V, Marcellin MW, Bilgin A (2017) The current role of image compression standards in medical imaging. Information. https://doi.org/10.3390/info8040131

Lu G, Fei B (2014) Medical hyperspectral imaging: a review. J Biomed Opt 19(1):10901. https://doi.org/10.1117/1.JBO.19.1.010901

Lu T, Liang P, Wu W-B, Xue J, Lei C-L, Li Y-Y, Sun Y-N, Liu F-Y (2012) Integration of the image-guided surgery toolkit (IGSTK) into the medical imaging interaction toolkit (MITK). J Digit Imaging 25(6):729–737. https://doi.org/10.1007/s10278-012-9477-3

Lyu Y, Jiang W, Lin Y, Voros L, Zhang M, Mueller B, Mychalczak B, Song Y (2018) Motion-blind blur removal for CT images with Wasserstein generative adversarial networks. In: 2018 11th international congress on image and signal processing, biomedical engineering and informatics (CISP-BMEI), pp 1–5. https://doi.org/10.1109/CISP-BMEI.2018.8633203

McFaline-Figueroa JR, Lee EQ (2018) Brain tumors. Am J Med 131(8):874–882. https://doi.org/10.1016/J.AMJMED.2017.12.039

Miao S, Wang ZJ, Liao R (2016) A CNN regression approach for real-time 2D/3D registration. IEEE Trans Med Imaging 35(5):1352–1363. https://doi.org/10.1109/TMI.2016.2521800

Miyakawa S, Tachibana H, Moriya S, Kurosawa T, Nishio T (2018) Evaluation of deformation parameters for deformable image registration-based ventilation imaging using an air-ventilating non-rigid phantom. Phys Med 50:20–25. https://doi.org/10.1016/j.ejmp.2018.05.016

Modat M, McClelland J, Ourselin S (2010) Lung registration using the NiftyReg package. In: Medical image analysis for the clinic: a grand challenge, workshop proc. from MICCAI 2010

Nam KW, Park J, Kim IY, Kim KG (2012) Application of stereo-imaging technology to medical field. Healthcare Inform Res 18(3):158–163. https://doi.org/10.4258/hir.2012.18.3.158

Nam JG, Park S, Hwang EJ, Lee JH, Jin K-N, Lim KY, Vu TH, Sohn JH, Hwang S, Goo JM, Park CM (2019) Development and validation of deep learning-based automatic detection algorithm for malignant pulmonary nodules on chest radiographs. Radiology 290(1):218–228. https://doi.org/10.1148/radiol.2018180237

Neumann B, Held M, Liebel U, Erfle H, Rogers P, Pepperkok R, Ellenberg J (2006) High-throughput RNAi screening by time-lapse imaging of live human cells. Nat Methods 3(5): 385–390. https://doi.org/10.1038/nmeth876

Nie D, Cao X, Gao Y, Wang L, Shen D (2016) Estimating CT image from MRI data using 3D fully convolutional networks. In: Deep learning and data labeling for medical applications: first international workshop, 2016, and second international workshop, DLMIA 2016, held in conjunction with MICCAI 2016, Athens, Greece, pp 170–178. https://doi.org/10.1007/978-3-319-46976-8_18

Nolden M, Zelzer S, Seitel A, Wald D, Müller M, Franz AM, Maleike D, Fangerau M, Baumhauer M, Maier-Hein L, Maier-Hein KH, Meinzer H-P, Wolf I (2013) The medical imaging interaction toolkit: challenges and advances: 10 years of open-source development. Int J Comput Assist Radiol Surg 8(4):607–620. https://doi.org/10.1007/s11548-013-0840-8

Oppermann S, Ylanko J, Shi Y, Hariharan S, Oakes CC, Brauer PM, Zúñiga-Pflücker JC, Leber B, Spaner DE, Andrews DW (2016) High-content screening identifies kinase inhibitors that overcome venetoclax resistance in activated CLL cells. Blood 128(7):934–947. https://doi.org/10.1182/blood-2015-12-687814

Panasyuk SV, Yang S, Faller DV, Ngo D, Lew RA, Freeman JE, Rogers AE (2007) Medical hyperspectral imaging to facilitate residual tumor identification during surgery. Cancer Biol Ther 6(3):439–446. https://doi.org/10.4161/cbt.6.3.4018

Panayides AS, Amini A, Filipovic ND et al (2020) AI in medical imaging informatics: current challenges and future directions. IEEE J Biomed Health Inform 24(7):1837–1857. https://doi.org/10.1109/JBHI.2020.2991043

Parker JR (2010) Algorithms for image processing and computer vision, 2nd edn. Wiley, Hoboken

Parsons AB, Lopez A, Givoni IE, Williams DE, Gray CA, Porter J, Chua G, Sopko R, Brost RL, Ho C-H, Wang J, Ketela T, Brenner C, Brill JA, Fernandez GE, Lorenz TC, Payne GS, Ishihara S, Ohya Y, Boone C (2006) Exploring the mode-of-action of bioactive compounds by chemical-genetic profiling in yeast. Cell 126(3):611–625. https://doi.org/10.1016/j.cell.2006.06.040

Payan A, Montana G (2015) Predicting Alzheimer's disease: a neuroimaging study with 3D convolutional neural networks. In: ICPRAM 2015—4th international conference on pattern recognition applications and methods, proceedings, 2

Penny W, Friston K, Ashburner J, Kiebel S, Nichols T (2007) Statistical parametric mapping: the analysis of functional brain images

Perlman ZE, Slack MD, Feng Y, Mitchison TJ, Wu LF, Altschuler SJ (2004) Multidimensional drug profiling by automated microscopy. Science 306(5699):1194–1198. https://doi.org/10.1126/science.1100709

Pettit C, Bishop I, Sposito V, Aurambout J-P, Sheth F (2012) Developing a multi-scale visualisation framework for use in climate change response. Landsc Ecol 27(4):487–508. https://doi.org/10.1007/s10980-012-9716-5

Pichler BJ, Wehrl HF, Kolb A, Judenhofer MS (2008) Positron emission tomography/magnetic resonance imaging: the next generation of multimodality imaging? Semin Nucl Med 38(3): 199–208. https://doi.org/10.1053/j.semnuclmed.2008.02.001

Pierce MC, Schwarz RA, Bhattar VS, Mondrik S, Williams MD, Lee JJ, Richards-Kortum R, Gillenwater AM (2012) Accuracy of in vivo multimodal optical imaging for detection of oral neoplasia. Cancer Prev Res (Phila) 5(6):801–809. https://doi.org/10.1158/1940-6207.CAPR-11-0555

Qasrawi R, Ivorra A (2015) Impact of liver vasculature on electric field distribution during electroporation treatments: an anatomically realistic numerical study. In: Lacković I, Vasic D (eds) 6th European conference of the international federation for medical and biological engineering. IFMBE proceedings, vol 45. Springer, Cham. https://doi.org/10.1007/978-3-319-11128-5_143

Qiu Y, Wang Y, Yan S, Tan M, Cheng S, Liu H, Zheng B (2016) An initial investigation on developing a new method to predict short-term breast cancer risk based on deep learning technology. In: Proc SPIE, pp 9785. https://doi.org/10.1117/12.2216275

Qiu D, Zhang S, Liu Y, Zhu J, Zheng L (2020) Super-resolution reconstruction of knee magnetic resonance imaging based on deep learning. Comput Methods Prog Biomed 187:105059. https://doi.org/10.1016/j.cmpb.2019.105059

Rajkomar A, Lingam S, Taylor AG, Blum M, Mongan J (2017) High-throughput classification of radiographs using deep convolutional neural networks. J Digit Imaging 30(1):95–101. https://doi.org/10.1007/s10278-016-9914-9

Ratib O, Ligier Y, Scherrer JR (1994) Digital image management and communication in medicine. Comput Med Imaging Graphics 18(2):73–84. https://doi.org/10.1016/0895-6111(94)90016-7

Rodríguez FJ, Staurenghi G, Gale R (2018) The role of OCT-A in retinal disease management. Graefes Arch Clin Exp Ophthalmol 256(11):2019–2026. https://doi.org/10.1007/s00417-018-4109-3

Scahill RI, Frost C, Jenkins R, Whitwell JL, Rossor MN, Fox NC (2003) A longitudinal study of brain volume changes in normal aging using serial registered magnetic resonance imaging. Arch Neurol 60(7):989–994. https://doi.org/10.1001/archneur.60.7.989

Schenone M, Dancik V, Wagner BK, Clemons PA (2013) Target identification and mechanism of action in chemical biology and drug discovery. Nat Chem Biol 9:232–240. https://doi.org/10.1038/nchembio.1199

Schirle M, Jenkins JL (2016) Identifying compound efficacy targets in phenotypic drug discovery. Drug Discov Today 21(1):82–89. https://doi.org/10.1016/j.drudis.2015.08.001

Seitel A, Yung K, Mersmann S, Kilgus T, Groch A, dos Santos TR, Franz AM, Nolden M, Meinzer H-P, Maier-Hein L (2012) MITK-ToF—range data within MITK. Int J Comput Assist Radiol Surg 7(1):87–96. https://doi.org/10.1007/s11548-011-0617-x

Shamir L, Delaney JD, Orlov N, Eckley DM, Goldberg IG (2010) Pattern recognition software and techniques for biological image analysis. PLoS Comput Biol 6(11):e1000974. https://doi.org/10.1371/journal.pcbi.1000974

Shi S, Yang R, You H (2017) A new two-dimensional Fourier transform algorithm based on image sparsity. In: 2017 IEEE international conference on acoustics, speech and signal processing (ICASSP), pp 1373–1377. https://doi.org/10.1109/ICASSP.2017.7952381

Shin H, Roberts K, Lu L, Demner-Fushman D, Yao J, Summers RM (2016) Learning to read chest X-rays: recurrent neural cascade model for automated image annotation. In: 2016 IEEE conference on computer vision and pattern recognition (CVPR), pp 2497–2506. https://doi.org/10.1109/CVPR.2016.274

Smith SM, Jenkinson M, Woolrich MW, Beckmann CF, Behrens TEJ, Johansen-Berg H, Bannister PR, De Luca M, Drobnjak I, Flitney DE, Niazy RK, Saunders J, Vickers J, Zhang Y, De Stefano N, Brady JM, Matthews PM (2004) Advances in functional and structural MR image analysis and implementation as FSL. Neuroimage 23(1):S208–S219. https://doi.org/10.1016/j.neuroimage.2004.07.051

Sone S, Kasuga T, Sakai F, Aoki J, Izuno I, Tanizaki Y, Shigeta H, Shibata K (1991) Development of a high-resolution digital tomosynthesis system and its clinical application. Radiographics 11(5):807–822. https://doi.org/10.1148/radiographics.11.5.1947318

Soni E, Mahajan R (2019) In: Luhach AK, Jat DS, Hawari KBG, Gao X-Z, Lingras P (eds) MRI image compression using asymmetric wavelet analysis BT—advanced informatics for computing research. Springer, Singapore, pp 260–270. https://doi.org/10.1007/978-981-15-0108-1_25

Sowell ER, Levitt J, Thompson PM, Holmes CJ, Blanton RE, Kornsand DS, Caplan R, McCracken J, Asarnow R, Toga AW (2000) Brain abnormalities in early-onset schizophrenia spectrum disorder observed with statistical parametric mapping of structural magnetic resonance images. Am J Psychiatry 157(9):1475–1484. https://doi.org/10.1176/appi.ajp.157.9.1475

Spampinato C, Palazzo S, Giordano D, Aldinucci M, Leonardi R (2017) Deep learning for automated skeletal bone age assessment in X-ray images. Med Image Anal 36:41–51. https://doi.org/10.1016/j.media.2016.10.010

Steelman ZA, Ho DS, Chu KK, Wax A (2019) Light scattering methods for tissue diagnosis. Optica 6(4):479–489. https://doi.org/10.1364/optica.6.000479

Sugawara Y, Tachibana H, Kadoya N, Kitamura N, Sawant A, Jingu K (2017) Prognostic factors associated with the accuracy of deformable image registration in lung cancer patients treated with stereotactic body radiotherapy. Med Dosimetry 42(4):326–333. https://doi.org/10.1016/j.meddos.2017.07.004

Suk H-I, Lee S-W, Shen D, Initiative ADN (2014) Hierarchical feature representation and multimodal fusion with deep learning for AD/MCI diagnosis. Neuroimage 101:569–582. https://doi.org/10.1016/j.neuroimage.2014.06.077

Teuwen J, Moriakov N (2020) Chapter 20—convolutional neural networks. In: Kevin Zhou S, Rueckert D, Fichtinger G (Eds.) Elsevier and MICCAI Society Book Series, Handbook of medical image computing and computer assisted intervention. Academic, pp 481–501

Tham NTT, Hwang S-R, Bang J-H, Yi H, Park Y-I, Kang S-J, Kang H-G, Kim Y-S, Ku H-O (2019) High-content analysis of in vitro hepatocyte injury induced by various hepatotoxicants. J Vet Sci 20(1):34–42. https://doi.org/10.4142/jvs.2019.20.1.34

Tripathy S, Swarnkar T (2020) Unified preprocessing and enhancement technique for mammogram images. Procedia Comp Sci 167:285–292. https://doi.org/10.1016/j.procs.2020.03.223

Uddin M, Wang Y, Woodbury-Smith M (2019) Artificial intelligence for precision medicine in neurodevelopmental disorders. NPJ Digit Med 2(1):112. https://doi.org/10.1038/s41746-019-0191-0

Wallyn J, Anton N, Akram S, Vandamme TF (2019) Biomedical imaging: principles, technologies, clinical aspects, contrast agents, limitations and future trends in nanomedicines. Pharm Res 36(6):78. https://doi.org/10.1007/s11095-019-2608-5

Wang LV, Wu HI, Masters BR (2008) Biomedical optics, principles and imaging. J Biomed Opt 13(4):049902. https://doi.org/10.1117/1.2976007

Wang Z, Lim EG, Tang Y, Leach M (2014) Medical applications of microwave imaging. Sci World J 2014:147016. https://doi.org/10.1155/2014/147016

Warchal SJ, Dawson JC, Carragher NO (2016) Development of the theta comparative cell scoring method to quantify diverse phenotypic responses between distinct cell types. Assay Drug Dev Technol 14(7):395–406. https://doi.org/10.1089/adt.2016.730

Wei J, Li G (2014) Automated lung segmentation and image quality assessment for clinical 3-D/4-D-computed tomography. IEEE J Transl Eng Health Med 2:1–10. https://doi.org/10.1109/JTEHM.2014.2381213

Wolfe WL (1997a) Introduction to imaging spectrometers. SPIE Press. https://doi.org/10.1117/3.263530

Wolfe W (1997b) Introduction to imaging spectrometers. Opt & Photonics News

Wolterink JM, Leiner T, de Vos BD, van Hamersvelt RW, Viergever MA, Išgum I (2016) Automatic coronary artery calcium scoring in cardiac CT angiography using paired convolutional neural networks. Med Image Anal 34:123–136. https://doi.org/10.1016/j.media.2016.04.004

Xizhi Z (2008) The application of wavelet transform in digital image processing. In: 2008 international conference on multimedia and information technology, pp 326–329. https://doi.org/10.1109/MMIT.2008.134

Yi X, Walia E, Babyn P (2019) Generative adversarial network in medical imaging: a review. Med Image Anal 58:101552. https://doi.org/10.1016/j.media.2019.101552

Yokoi T, Suzuki K, Oba K (1991) Ultraviolet light imaging technology and applications. Proc SPIE, p 1449. https://doi.org/10.1117/12.44263

Yoon HJ, Jeong YJ, Kang H, Jeong JE, Kang D-Y (2019) Medical image analysis using artificial intelligence. Progress Med Phys 30(2):49–58. https://doi.org/10.14316/pmp.2019.30.2.49

Young DW, Bender A, Hoyt J, McWhinnie E, Chirn G-W, Tao CY, Tallarico JA, Labow M, Jenkins JL, Mitchison TJ, Feng Y (2008) Integrating high-content screening and ligand-target prediction to identify mechanism of action. Nat Chem Biol 4(1):59–68. https://doi.org/10.1038/nchembio.2007.53

Yushkevich PA, Piven J, Hazlett HC, Smith RG, Ho S, Gee JC, Gerig G (2006) User-guided 3D active contour segmentation of anatomical structures: significantly improved efficiency and reliability. Neuroimage 31(3):1116–1128. https://doi.org/10.1016/J.NEUROIMAGE.2006.01.015

Zanfardino M, Franzese M, Pane K, Cavaliere C, Monti S, Esposito G, Salvatore M, Aiello M (2019) Bringing radiomics into a multi-omics framework for a comprehensive genotype–phenotype characterization of oncological diseases. J Transl Med 17(1):337. https://doi.org/10.1186/s12967-019-2073-2

Zhang C, Shen X, Cheng H, Qian Q (2019) Brain tumor segmentation based on hybrid clustering and morphological operations. Int J Biomed Imaging, 7305832. https://doi.org/10.1155/2019/7305832

Zhang Y, Wu X, He L, Meng C, Du S, Bao J, Zheng Y (2020) Applications of hyperspectral imaging in the detection and diagnosis of solid tumors. Transl Cancer Res 9(2):1265–1277. https://doi.org/10.21037/tcr.2019.12.53

Zhang J, Shang Y, Li X, Dong Y, Pei Y (2021) Novel technique for full-field deformation and temperature measurement by ultraviolet imaging: experimental design and preliminary results. Coatings 11(6):641. https://doi.org/10.3390/coatings11060641

Zonios G, Perelman LT, Backman V, Manoharan R, Fitzmaurice M, Van Dam J, Feld MS (1999) Diffuse reflectance spectroscopy of human adenomatous colon polyps in vivo. Appl Opt 38(31):6628–6637. https://doi.org/10.1364/ao.38.006628

Artificial Intelligence and Its Application in Cardiovascular Disease Management

9

Vigneshwaran Namasivayam, Nithyashree Senguttuvan, Venkatesan Saravanan, Senthilkumar Palaniappan, and Muthu Kumaradoss Kathiravan

Abstract

Artificial intelligence (AI) is a type of computing technology that enables machines to perform cognitive functions like a human brain. In comparison to manual or traditional approaches, AI can provide improved results and efficient outcomes with minimal errors. Machine learning and deep learning trains to act or react according to the data and results in an efficient output. Several types of AI algorithms (supervised or unsupervised) lead to significant improvement in prediction accuracy for various application of science. The efficient storage of

Vigneshwaran Namasivayam and Nithyashree Senguttavan contributed equally with all other contributors.

V. Namasivayam
Pharmaceutical Chemistry, Pharmaceutical Institute, University of Bonn, Bonn, Germany

N. Senguttuvan
Department of Computer Science, Sri Krishna College of Technology, Kovaipudur, Tamil Nadu, India

V. Saravanan
Dr. APJ Abdul Kalam Research Lab, SRM College of Pharmacy, SRMIST, Kattankulathur, Tamil Nadu, India

S. Palaniappan (✉)
Faculty of Pharmacy, Karpagam Academy of Higher Education, Coimbatore, Tamil Nadu, India
e-mail: drsenthilkumar.p@kahedu.edu.in

M. K. Kathiravan (✉)
Dr. APJ Abdul Kalam Research Lab, SRM College of Pharmacy, SRMIST, Kattankulathur, Tamil Nadu, India

Department of Pharmaceutical Chemistry, SRM College of Pharmacy, SRMIST, Kattankulathur, Tamil Nadu, India
e-mail: kathirak@srmist.edu.in

S. Singh (ed.), *Machine Learning and Systems Biology in Genomics and Health*,
https://doi.org/10.1007/978-981-16-5993-5_9

patient-related information, analysis, diagnosing diseases has triggered the growth of AI and its application in the medical field. Pioneers from various domains have successfully developed and launched AI algorithms and developed tools in the healthcare domain. Particularly in cardiovascular disease management, AI finds its application in precision, predictive, and preventive medicine increasing the efficiency of healthcare professionals and enabling them for better diagnosis and treatment. Recently, FDA has authorized AI-based medical devices for healthcare management.

9.1 Introduction

Technology is an integral part of our life. Wherever we are, we are accompanied with technology as a companion. People have been replaced with technology to live with. As the expectations are increasing day by day, it had marked itself a standard of development from its early stage. From the invention of the wheel in the past to the autonomous machines and robots now, it had established a form of necessity and technology has earned a place in the world. Future technology which is in accordance with the term "emerging technology" changes with the timeline running in past, present, and future. With the evolving trend on technology, people no more update technology for their needs, it updates accordingly with the implementation in various fields. The different areas of its applicability are broadcasting, occupational, information, or education sector. The technologies that are under or in the near future development can be termed "emerging technology." The current technological trends are artificial intelligence (AI), machine learning (ML), virtual reality, augmented reality, robotic process automation, edge computing, Internet of Things, 5G, quantum computing, block chain, and cyber security. These technologies along with the necessary professional experience increase job opportunities.

The future of healthcare lies in the application of technology by healthcare professionals, as these technologies need support in order to stay relevant to the domain. These unsustainable ones are being made into sustainable ones to bring up faster, cheaper, and effective solutions to combat diseases. Technologies have fought against AIDS, cancer, Ebola, or COVID-19 and had saved our lives by making our communities healthy (Dagnino 2021).

9.2 Artificial Intelligence

AI is an innovation in the replication of human intelligence in machine form. AI requires customized hardware and software to develop and train ML algorithms (Fig. 9.1) using programming languages like Python, R, and Java. This system works by considering labeled training data, analyzing it for correlations, recognizing

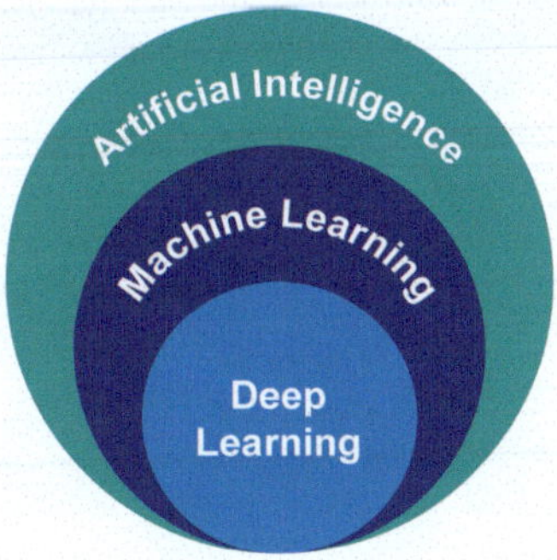

Fig. 9.1 Different hierarchies of artificial intelligence (AI)

patterns, and making predictions. AI focuses on three cognitive skills: **learning**—acquire data for developing algorithms, **reasoning**—to obtain the desired outcome by selecting the correct algorithm, and **self-correction**—continuously optimizing the algorithms. Some applications of AI are machine vision, speech recognition, natural language processing, and expert systems (Pannala et al. 2020).

AI is important as it can provide insights into tasks that are yet to be explored. They provide improved results than humans in repetitive, detail-oriented tasks and often finish jobs promptly with relatively minor errors. The advantages of AI are that it is available at any time of need, reduces time and work, and the precise outcome.

9.3 Market Valuation of AI

The market value of AI was globally calculated as USD 62.35 billion (2020) and is projected to increase at a compound annual growth rate (CAGR) of 40.2% (2021–2028). AI is the technological advancement of computing systems that are capable of performing tasks that include decision-making, speech recognition, visual perception, and language translation. It utilizes multiple layers of algorithms to understand, visualize, and process information that are involved in data processing, calculation, and automated reasoning. The algorithms are improved on a daily aspect, as conventional algorithms are doubted on their accuracy and efficiency which had led to the development of standard algorithms. The continuous research and innovation by various industry rulers had bought in advanced technologies in the area of automotive, healthcare, retail, finance, and manufacturing. AI has owned its place with self-driving vehicles to crucial life-saving medical gear by being infused virtually into every machine and program. Leading technology companies like Amazon, Google, Apple, Facebook, International Business Machines Corporation (IBM), and Microsoft are capitalizing on AI and are working intensively to create AI more accessible for enterprise applications (Steinhoff 2021).

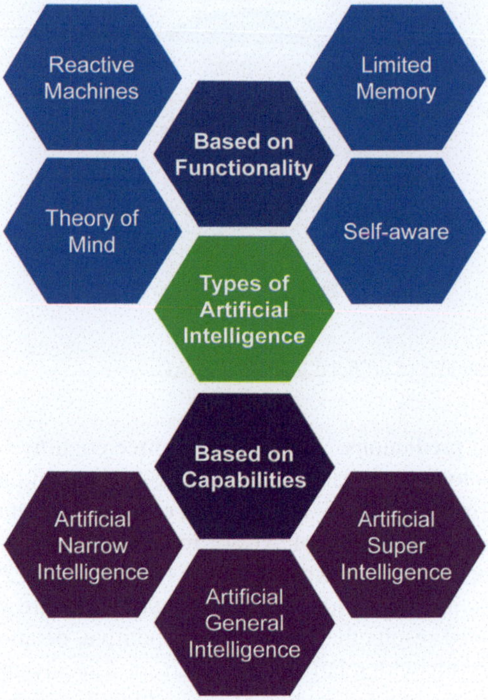

Fig. 9.2 Different types of artificial intelligence

9.4 Typology of AI

AI was created to match the functions of machines with humans to the degree where it can replicate human capabilities. Thus, based on the comparison in terms of adaptability and performance, AI can be categorized into many types. An evolved type of AI is that it can execute human-like tasks with comparable levels of proficiency or alternatively as less evolved type AI which has limited functionality and performance (Paschen et al. 2020) (Fig. 9.2).

9.4.1 Classification Based on Functionality

Based on the functional ability of AI to "think" and perhaps even "feel" like humans, AI can be grouped into four types—reactive machines, limited memory machines, theory of mind, and self-aware AI.

9.4.1.1 Reactive Machines

Reactive machines are the oldest form of AI which have a restricted capacity and can reproduce the skills of the human brain to respond to stimuli. Since the reactive machines lack memory-based functionality, previously gained experiences cannot be used to respond for present actions. They are used in responding to a limited set or combination of inputs. A well-known application of a reactive machine is IBM's Deep Blue, which beat chess Grandmaster Garry Kasparov in 1997 (Nilsson and Nilsson 1998).

9.4.1.2 Limited Memory

Limited memory machines possess the capability of training from past information to arrive decisions along with the abilities of reactive machines. All present-day deep learning (DL) AI systems learn with the help of a large amount of training data which are stored in their memory and used as a reference model to solve similar future problems. For example, an image recognition applying AI is trained with thousands of pictures with labels so that it can name the objects it scans. Chatbots, virtual assistants, and self-driving cars are few technologies driven by limited memory AI. The three major kinds of machine learning models in limited memory are:

Reinforcement Learning: Acquire information through the sequence of trial and error and the model is used to train computers to play games such as chess, Go, and DOTA2 (Barto and Sutton 1997)

Long Short-Term Memory (LSTMs): The historical data helps to predict the next item in sequences and, in particular, they are widely beneficial in language. The LSTM tags the information based on their occurrence (recent or past) and predicts the next elements in the sequence

Evolutionary Generative Adversarial Networks (E-GAN): The memory which evolves at every evolution and finds an optimal path varied due to various modifications. The model can identify a better and less resistance path by these modifications such that the next generation of the model modifies and evolves, where there was an error in the previous path. Like humans evolved on this planet by successfully multiplying and survive a life better than its parent, similarly, the E-GAN creates a simulation

9.4.1.3 Theory of Mind

A theory of mind is the next level of AI where the researchers are currently working on understanding the things which are interacting with by discerning their needs, emotions, beliefs, and thought processes. To perceive human needs, the individual thought process of every human should be considered. This process requires more understanding and is yet to be in the application.

9.4.1.4 Self-Aware

Self-aware AI is a hypothesis that has established self-awareness like humans. This AI can be created in the near or long future by materializing and is the main aim of all AI research. The self-aware AI can understand and induce emotions in those it communicates with, and also have emotions, needs, beliefs, and desires of its own. Moreover, many believe that this is the state where the end of humanity can happen as these characteristics of AI can stimulate self-preservation.

9.4.2 Classification Based on Capability

Another classification system of AI is artificial narrow intelligence (ANI), artificial general intelligence (AGI), and artificial superintelligence (ASI).

9.4.2.1 Artificial Narrow Intelligence (ANI)

Artificial narrow intelligence, also known as narrow AI or weak AI, is an AI system that can accomplish specific tasks independently using human-like capabilities. The machines follow their programs for actions and have a very thin range of competencies. They resemble all the reactive and limited memory AI. ML and DL are part of this category.

9.4.2.2 Artificial General Intelligence (AGI)

Artificial general intelligence is the capability of an AI to replicate human-like multifunctional capabilities. These systems need less time for training and can autonomously build several competencies, connections, and generalizations across domains.

9.4.2.3 Artificial Super Intelligence (ASI)

The development of artificial super intelligence will become by far the most skilled form of intelligence on the planet due to superior memory, quick data processing and analysis, and decision-making abilities; it will be better at everything along with the multi-faced human-like intelligence. These machines have the potential power to threaten human existence (Girasa 2020).

9.5 Evolution of AI

In 1642, Pascal's development of the first mechanical calculating machine opened the stage for the rise of machine dominance. Later in 1837, Charles Babbage, the father of the computer, offered the first concept as a programmable machine. In 1800s, AI was restricted in myths, fiction, and speculations. Traditional philosophers anticipated machines integrated into human beings which were represented in fiction work like Mary Shelley's "Frankenstein." During the first half of the twentieth century, various science fiction included the concept of robots like the "heartless" Tin man from the *Wizard of Oz* and the humanoid robot that impersonated Maria in *Metropolis*.

As a stepping-stone in the evolution of modern AI, in 1943, McCulloch and Pitts established the foundation of neural networks synchronizing between the brain and computing machines. By 1950s, many scientists, mathematicians, and philosophers introduced the concept of AI (Fig. 9.3). Notable among them is Alan Turing, a young

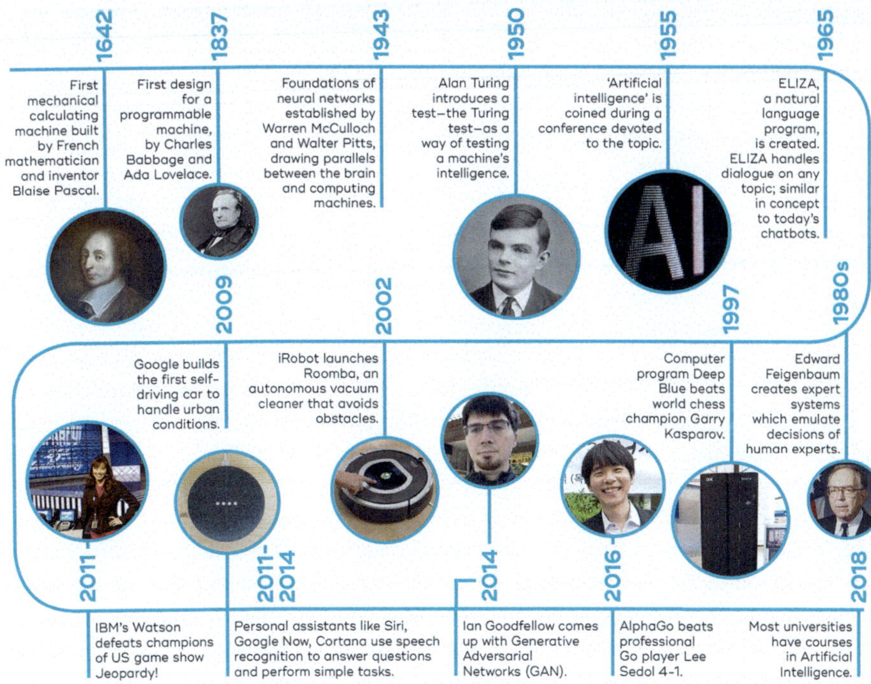

Fig. 9.3 History of AI (Courtesy: the University of Queensland, Australia (History of Artificial Intelligence n.d.))

British polymath, who proposed that machines can also use information to solve problems and make decisions. The paper discusses different ways to build intelligent machines and validate their intelligence.

The real beginning of AI was initiated in 1956 in a workshop at Dartmouth College. The term "proof of concept" was initialized through the program "Logic Theorist" by Allen Newell, Cliff Shaw, and Herbert Simon, which mimics the problem-solving skills of a human. It is considered as the first artificial intelligence program and was presented at the *Dartmouth Summer Research Project on Artificial Intelligence* hosted by John McCarthy and Marvin Minsky.

From 1957 to 1974, AI had an excellent improvement as computers managed to store more data, became faster, cost-effective, and easily accessible. ML algorithms are explored by applying algorithms to individual problems of the people. Newell, Shaw, and Simon's *General Problem Solver* (1959), a problem-solving tool, and Joseph Weizenbaum's *ELIZA* (1965), a natural language program created to discuss various topics like chatbots, were developed. The successful outcomes of these projects persuaded the government agency, Defense Advanced Research Projects Agency, to support AI research involving transcription and translation of spoken language with high-throughput data processing.

In further discoveries of AI, there was a lack of computational power, memory to store data, and processing speed. In 1980s, AI gave its comeback with two sources:

an expansion of the algorithmic toolkit and enhanced funds. John Hopfield and David Rumelhart popularized DL techniques allowing computers to learn using past datasets. Similarly, Edward Feigenbaum introduced expert systems that made decisions like humans. For every situation, an expert in that field would be asked for their decision, and later these were used to give advice to nonexperts. These were widely used in industries.

In the period between 1990 and 2000, many of the milestones of AI had been accomplished. In 1997, IBM's *Deep Blue*, a chess-playing computer program, defeated Gary Kasparov and this was the first time a reigning world chess champion and grandmaster lost to a computer. This was publicized widely and was a major step towards an AI decision-making program. In the same year, Microsoft Windows implemented speech recognition software developed by Dragon Systems. It appeared that machines were capable of handling anything including human emotions like Kismet, developed by Cynthia Breazeal, which recognized and displayed emotions (Husbands et al. 1997; McNally et al. 2012).

AI's evolution in the twentieth century was inevitable. In 2009, the Graphical Processing Unit (GPU) was discovered and played a major role in the development of AI. This development provides an edge for various AI-based platforms. In the same year, Google has incorporated AI in the automobile industry for autonomous vehicles. In mid-2015, the introduction of chatbots improved customer care services. In 2017, the Arterys, a cloud-based DL application in the healthcare sector, was the first to be approved by FDA, and later AI-based esophagogastroduodenoscopy platform started trials in gastroenterology. Further in 2019, Siemens Medical solutions introduced its one-of-a-kind product for the cardiovascular system, AI-RAD Companion. Following this, many AI-based platforms and algorithms were under approval. Recent inventions such as SpayaAI in retrosynthesis and Google AI DeepMind program AlphaFold predicts protein structure encourages the application of AI in drug discovery. The advancement in science and technology had led to the evolution of AI (Fig. 9.4), and in particular the developments in healthcare are highlighted.

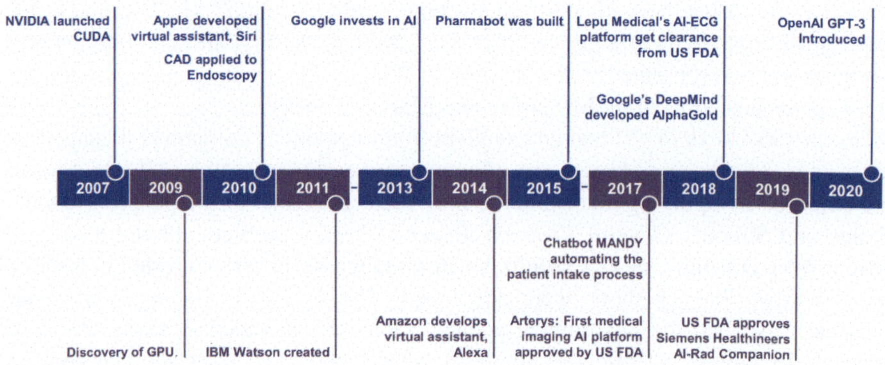

Fig. 9.4 Evolution of AI

9.6 Applications

The main purpose of AI is to help humans with human capabilities to make advanced decisions with better consequences. From a philosophical perspective, AI makes humans life meaningful, free from hard labor, and help to manage the complex connection of individuals, companies, states, and nations to function in a beneficial way.

Currently, AI is employed to improve process efficiencies, automate resource-heavy tasks, and make business predictions. The research and development cost related to the system components of AI must be subsidized by the corporations and government agencies for enhancing the development of AI in different applications. AI is used to give perceptions and proposals on the basis of information from various domains. For example, Google's predictive search algorithm predicts the input from the user in the search bar. Netflix recommends a movie to its users, engaging them in the platform and increasing the watch time. Facebook provides suggestions to tag friends on their photos, based on their facial features in the images.

The major applications in our everyday life are website predictions, maps, car sharing, autopilot, spam filters, plagiarism identifiers, face and voice recognition, online shopping platform search suggestions, fraud protection/prevention, and personal assistants such as Siri, Alexa, and Bixby. AI-based tools are applied significantly in healthcare such as MYCIN is used to detect microbes and prescribe personalized antibiotic therapy. IBM developed cloud-based AI technology for lung cancer patients to provide assistance in the treatment. Intel had aided in accelerating magnetic resonance imaging (MRI), genetic research, radiology, AI-powered disinfection robots, and inpatient monitoring technology. Google health is enabled with AI to detect disease, data storage, and enabling providers with hardware and insurance (Strong 2016). The application of AI in different fields is summarized in Table 9.1.

9.6.1 Healthcare

Healthcare is the collective team effort for maintaining and or improving the quality of life via health by preventing and treating the physical and mental impairment of human beings. Healthcare is serviced by the medical, health professionals, and allied fields in health sciences. It generally includes primary, secondary, tertiary, and public healthcare. Primary care is generally the direct access to the health visitors, whereas secondary care includes services with more specialist knowledge. Tertiary care is a highly support specialist and required in a complex condition with specialist hospitals. For the well-being of human life with the advancement of science, the World Health Organization provides great efforts to give everyone, everywhere an equal chance to live a healthy life (Hamet and Tremblay 2017; Ramesh et al. 2004). In recent times AI has found a wide range of applications in the healthcare sector starting from administrative tasks, diagnosis, surgery, and monitoring of patients.

Table 9.1 Application of AI in different fields

Field	Department	Application/advantages
E-commerce	Recommendations	– Customer service in an efficient way
	Chat bots	– Any time accessibility
	Filtration	– Predict spam and fake reviews
	Optimizing search	– Suggest to find the right product
	Supply chain	– Predicts demand to ensure stocks are available for the supply
Human resources	Work culture	– Employee data analysis – Team management – Project assignment – Feedback analysis – Commitment of employees
	Hiring	– Skillset analysis
Robots	Assembly	– Real-time course correction (along with advanced vision systems) – Helps to learn the best path for certain process in operation
	Customer service	– Customer interaction – Problem-solving
	Packaging	– Quicker, cheaper, and accurate packaging – Robots can refine motions at ease by learning
	Open-source	– Perform custom tasks for a specific application
Education	Grading	– Student assessment – Analysis of student's requirement
	Tutor	– Guidance to students
Finance	Applications	– Provide financial advice – Intuit mint or TurboTax
	Trading	– Trading on wall street
Law	Document searching	– Looks through documents at a faster rate
	Analysis	– Machine learning understands data and predicts outcomes – Computer vision for classification and extraction of data – Natural language processing to understand requests
Banking	Chat bots	– Make the customers familiarize bank services and carry out transactions
	Virtual assistant	– Improves and cost cutting of compliance with regulations
	Taking decisions	– Helps to choose loans, set the limits for credits, and find out new investments
Transportation	Management	– Operates autonomous vehicles, air and road traffic control, and safe shipping
Security	Malware detection	– Detection of anomalies
	Analysis	– Identifying malicious code and cyber-attacks

- **Administration:** AI systems have been widely used in routine, daily tasks and to reduce human mistakes and improve efficiency. AI helps to transcript medical records using natural language processing (NLP) and organizes the information of patients for easy access to doctors.
- **Telemedicine:** In nonemergency cases, patients can get in touch with the AI system in the hospital for investigating their vital symptoms and need for medical assistance, thereby saving the time of physicians and healthcare professionals.
- **Assisted Diagnosis:** ML is being utilized by the healthcare system for faster and better diagnosis. AI can read MRI scans for tumors and other malignant growths in a faster way with minimum error in comparison to humans. An array of AI technologies is used to understand, predict, and combat deadly infectious diseases such as HIV, Ebola, and COVID-19.
- **Robot-Assisted Surgery:** The surgeries using robotic assistance have minimal error and can function continuously. The degree of accuracy of the robot's function is so high that the surgeries are less invasive than the traditional methods, thereby reducing the hospital recovering period for patients.
- **Vital Stats Monitoring:** Body vital stats are available on tap nowadays with the advent of wearable devices achieving mass-market popularity. This can predict the health deteriorations even before the patient recognizes and saves human life (Jiang et al. 2017; Haleem et al. 2019; Yu et al. 2018).

9.7 Artificial Intelligence

9.7.1 Machine Learning

Machine learning (ML) performs tasks autonomously without being explicitly programmed. IBM has been connected with ML for a long time. Arthur Samuel, an employee of IBM then, invented the term "machine learning" while researching the game of checkers program. ML is a process where computers learn from the input data for carrying out a certain task. Algorithms involving the steps necessary for assigning simple tasks to the computers can be easily programmed without the learning needed by the computer. However, for advanced tasks, creating an algorithm manually is a bit complicated and challenging for a programmer. Rather, if the machine is allowed to create its own task list for performing the steps needed, the results would be more effective. ML uses different methodologies to teach computers for completing tasks where there is difficulty in manually creating or unavailability of satisfactory algorithms, which can further be used as training data to improve the algorithm by the computers. One of the classical examples is the MNIST dataset which stores the handwritten digits utilized for digital character recognition.

ML is a subclassification of AI-based on the fact that machines can train from available data to recognize patterns and provide conclusions with less human contribution. It was developed from pattern recognition, and machines can learn without being explicitly programmed. The learning process is started from observation or feeding of information in search of patterns and provides a better decision.

This process is iterative for generating a better model and identifying reliable outcome. An example of an ML application is AlphaFold created by DeepMind, one of Google's ventures that has figured out the solution for protein folding problems (Jordan and Mitchell 2015).

ML consists of the following three tasks involved in it.

1. **Decision Process:** The algorithm estimates the pattern in the input data.
2. **Error Function:** Estimates the model prediction, and in the case of known examples, comparison is made to assess the accuracy of the model.
3. **Model Optimization Process:** The parameters are modified to minimize the discrepancy between the known example and the model prediction. Optimization will continue until certain accuracy of the model is reached.

9.7.2 Deep Learning

Deep learning (DL) reflects the functioning of the human brain, thereby creating patterns to help in making a better decision. DL is a subsection of ML that has the capability to learn from unorganized data. DL uses neural networks with computing power to understand patterns from huge data. They are good at recognizing objects in pictures and words in sounds. They are now being employed in various tasks including self-translation of languages, diagnosis in the medical field, and playing a key role in socioeconomic problems.

The word "deep" refers to the multiple layers in the network with respect to DL. A network with a non-polynomial activation function and a hidden layer can be a universal classifier while linear perceptron could not be achieved. It is considered a modern variation with an unbounded number of layers of bounded size and allows practical application with optimized implementation. The heterogeneous layers make it more efficient, trainable, and understandable while being structured. Thus, it employs a hierarchical level of artificial neural networks using ML leading to the construction of the human brain, with neuron nodes similar to a web. The classical programs construct data in a linear way whereas the hierarchical function of DL approaches uses the nonlinear model. The architecture of DL comprises deep neural networks, deep belief networks, deep reinforcement learning, recurrent neural networks, and convolutional neural networks. They are utilized in various areas such as computer vision, speech recognition, natural language processing, machine translation, bioinformatics, drug discovery, medical image analysis, and material analysis (Esteva et al. 2019; El-Amir and Hamdy 2020).

The learning in AI can be classified into supervised and unsupervised learning. The supervised learning is referred to a system trained using labeled data which are divided into classes and assigning test data to these classes. The two categories of supervised learning are classification and regression. In unsupervised classification, the system uses unlabeled data and attempts to find similarities in order to build clusters. The unsupervised learning is categorized into clustering, anomaly detection, and dimensionality reduction (Fig. 9.5).

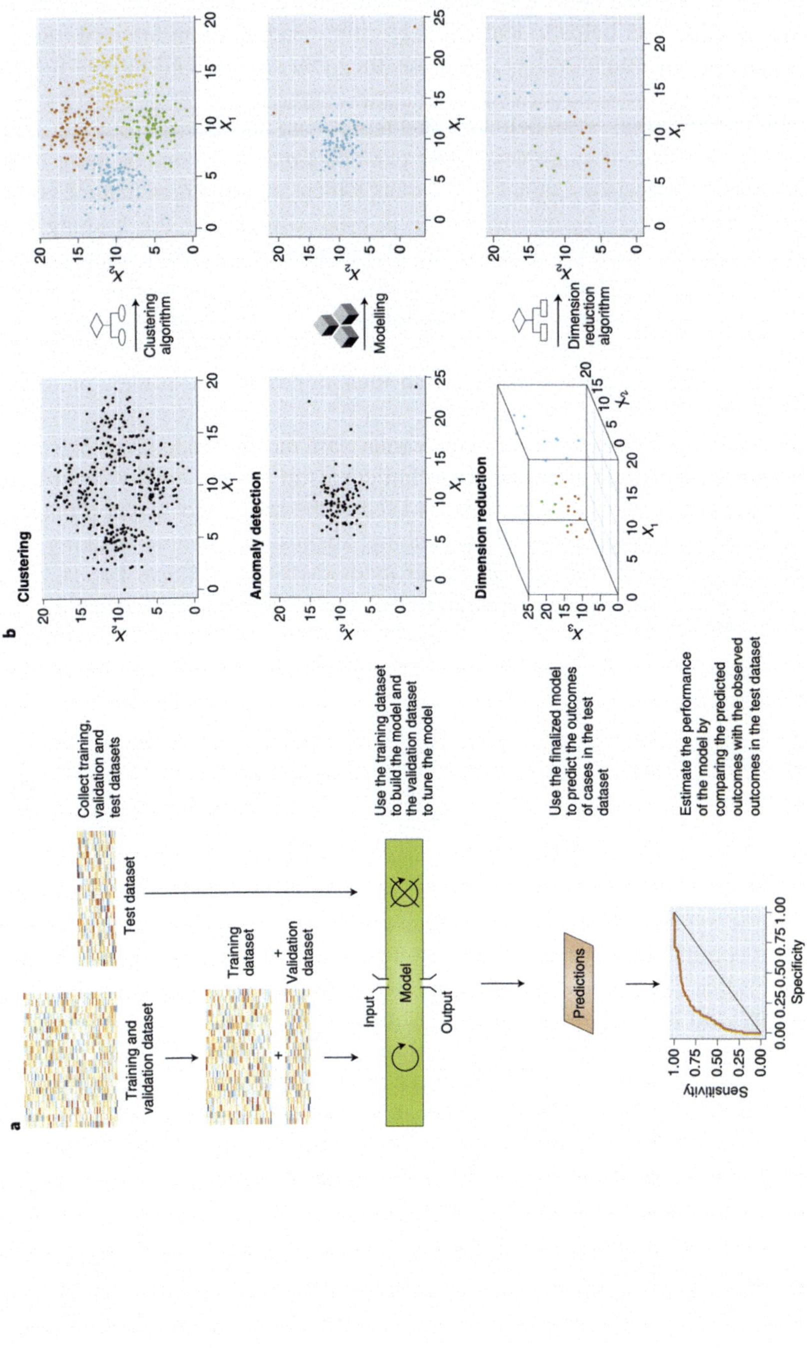

Fig. 9.5 (**a**) Supervised and (**b**) Unsupervised learning

9.7.2.1 Supervised Learning

Supervised learning algorithms are practiced with labeled examples where the desired output is known. Based on the input, the algorithm learns by comparing its actual output and identifies its errors by which it modifies the model. It can also be applied by using historical data (known data) and predict future events. The training from the known dataset helps to learn the algorithm leading to an inferred function for the prediction of output values, and thus they can recommend targets for new input once trained.

The algorithms build a model with the inputs and the desired outputs using training data with a set of examples. These examples may have one or more inputs which are called supervisory signal. The inputs which are known as feature vectors are filled in an array or vector and matrix is used to represent the training data. Weights are adjusted when the data is being filled into the model to fit appropriately. This is due to the cross-validation process which ensures that the model does not lead to overfit or underfit. Through iterative optimization, the algorithm learns to predict the output and an optimal function determines the output of test data. An algorithm is considered to be learning the task based on their improvement in their prediction accuracy.

The different categories of supervised learning are mostly made up of classification and regression. A classification algorithm comes in when there are constrained outputs, whereas a regression algorithm is used if the output possesses a range of numerical values. Applications include ranking, suggestion, identity tracking, face identification, and sound verification. Other methods used are Naïve Bayes, linear regression, logistic regression, random forest, and support vector machine. DL removes feature engineering by converting the data into a representation similar to principal components and leads to layered structures removing the unnecessary features in the representation.

Artificial Neural Network

Artificial neural networks (ANNs) are computing systems vaguely inspired by biological neural networks. They "learn" to perform tasks using examples on its own. The model is based on connected units or nodes named "artificial neurons," similar to neurons in the brain. The connection can transmit information, a "signal," similar to the synapse in the brain. The signal is processed and is sent to other connected neurons. An edge is a connection between the neurons. These neurons have weights that can fluctuate based on the signal strength in a connection. Artificial neurons can send a signal only when the aggregate signal surpasses the threshold. The neurons are aggregated into layers performing various types of transformations based on their inputs. From the input layer to the output layer signals travel by traversing the layers many times. This was all to meet the original goal of the ANN similar to human brain function. This changed with time as it is now being employed in computer vision, speech recognition, machine translation, filtering of social media content, playing games, and medical diagnosis.

Similar to the human brain, DL processes light as well as sound and converts into vision and hearing. They are comprised like ANNs with multiple hidden layers.

Neurons may have states, represented by real numbers in the range from 0 to 1. The weights of neurons and synapses change as learning proceeds depending upon the strength of the signal. A neural network greater than three layers is said to be a DL algorithm or a deep neural network, and a neural network less than that is a basic neural network. Neurons are organized in multiple layers with each layer performing transformations on every input. The signals travel from the input transverse through multiple layers to reach the output layer. AI has helped in tasks difficult to express using the classical algorithm. As of 2017, neural networks possess thousand to a few million units with millions of connections. Though the order of magnitude is less when compared to the human brain, still they perform tasks beyond human brainpower.

Recurrent Neural Network

A recurrent neural network (RNN) is a DL artificial neural network that uses sequential data or time-series data. They are generally used for problems like language translation, NLP, speech recognition, and image captioning. It is derived from feed-forward neural networks, and RNNs can use their internal memory to process variable-length sequences of inputs. This makes them applicable to tasks such as handwriting recognition or speech recognition. Many applications like Siri, voice search, and Google Translate use this.

They also use trained examples for learning and are distinguished by their "memory" because they gather information from prior and future inputs to look into the current input and output. Moreover, unidirectional recurrent neural networks cannot use future events for their predictions. The salient characteristics of RNN are the hidden state which remembers the sequence. RNN uses the memory which reduces the complexity of parameters and utilizes convolutional layers to extend the effective pixel neighborhood.

The term "recurrent neural network" refers to two broad classes having a similar general structure with temporal dynamic behavior. The first one is a finite impulse with a directed acyclic graph that can be unrolled, and the other is an infinite impulse with a directed cyclic graph that cannot be unrolled. Both of the impulses can have additional stored states under the control by the neural network which can be replaced by other networks if possess time delays or has feedback loops. The specific controlled states are referred to as gated state or gated memory and are part of LSTMs network and gated recurrent units which are also known as feedback neural network. The RNNs are made deep with the addition of hidden states or nonlinear hidden layers.

1. **Bidirectional RNNs**
   In speech recognition and handwriting recognition tasks, Bidirectional RNNs are combined with NLP, and the inputs tend to be available for an entire sentence which can be considered at once. This model can be customized based on application.
2. **Recursive Neural Networks**
   A recursive neural network need not apply inputs in a sequential fashion differing from recurrent neural networks based on linear models.
3. **Encoder-Decoder Sequence to Sequence RNNs**

Encoder-decoder sequence to sequence RNNs is applied in translation services with two RNNs. An encoder updates the hidden state producing "context" output fed to the decoder which translates the outputs depending upon the length.

4. **Long Short-term Memory**
   "LSTMs" are a special kind of RNN, proficient in learning long-term dependencies. Hochreiter and Schmidhuber in 1997 introduced them and in subsequent years it was optimized by many people. The main advantage is that they can work well on a variety of problems and hence it is often applied to many problems. A typical LSTM unit possesses a cell, an input, and an output gate with a forget gate. The cell registers the values over arbitrary time intervals and the three gates control the flow of information into and out of the cell. This flow is carried out by sigmoid and tanh functions in the various gates.

   LSTM networks are well suited to classify, process, and make predictions depending upon time series data because there can be unknown duration lags between key events in a time series. They deal with the vanishing gradient problem found in training traditional RNNs. The main advantage of LSTM over other models is relative insensitivity to gap length.

   LSTMs in general were explicitly designed to avoid the long-term dependency problem. All the RNNs possess chain form with repeat modules of a neural network and have a simple structure like a single tanh layer. This chain form is present in LSTMs but the repeat module has a different structure. The single neural network layer is replaced with four for the purpose of interaction.

   LSTM variant, introduced by Gers and Schmidhuber (2000), is adding "peep-hole connections" with gate layers looking at cell state. Another variation is to use coupled forget and input gates. Instead of separate decision-making, they are generally coupled together such as what to forget and what to add up. There are lots of others, such as Depth Gated RNNs by Yao, et al. reported in 2015 and approached to tackle long-term dependencies like Clockwork RNNs by Koutnik, et al. in 2014. LSTM in general introduces changes in computation of outputs and hidden state using the inputs and improves simple transformation by introducing additional gates and a cell state.
5. **Gated Recurrent**
   The gated recurrent unit, or GRU, is a version of LSTM introduced by Cho et al. (2014). It combines forget and input gates into a single "update gate" with the merge of cell state and hidden state. The resulting model is simpler which has increased popularity. The GRU is similar to an LSTM but a new version of a recurrent neural network with two gates, a reset gate and an update gate. They are a bit faster when compared to LSTMs. Researchers and engineers employ both for their application and choose the best one.

Convolution Neural Network

A convolutional neural network (CNN) is a strong image processing AI using generative and descriptive tasks. They include image and video recognition, along with recommender systems and NLP. Traditional neural networks cannot perform image processing and must be provided with reduced-resolution pieces. The neurons of CNN are arranged to overcome this by arranging them as like the frontal lobe that processes visual stimuli.

A CNN is like a multilayer perceptron with an input layer, an output layer, and a hidden layer that includes multiple convolutional layers, pooling layers, fully connected layers, and normalization layers. These are dedicated for image and natural language processing. ConvNets have the ability to learn filters/characteristics on training and the built is similar to neuron patterns in the human brain inspired from visual cortex. Individual neurons respond to stimuli only in a restricted region of the visual field known as the receptive field and a group of such fields covers the area of the image it has been visualized.

The ConvNet uses relevant filters to capture the spatial and temporal dependencies in an image. The architecture performs a better fitting to the image dataset due to the reduction in the number of parameters involved and the reusability of weights. The convolution operation extracts high-level features such as edges and is not limited to only one convolutional layer. The first ConvLayer is responsible for capturing the low-level features, and further the architecture adapts to the high-level features providing a better understanding of images in the dataset.

The results are of two types wherein one the convolved feature is reduced in dimensionality by valid padding and the other increases or has the same dimensionality as of input by the same padding. The pooling layer is also responsible for reducing the spatial size of the convolved feature and is further used for extracting the dominant features which maintain the process of training the model effectively. There are two types of pooling: max pooling to return maximum values and average pooling to return the average of all the values. Max pooling also performs as a noise suppressant where it discards the noisy activations. The model after being learned produces final output, that needs to be flattened to feed.

A fully connected layer is generally a cheap way of learning nonlinear combinations. The image suitable for multilevel perceptron is flattened into a column vector. This is fed to a feed-forward neural network where backpropagation is applied to iterations of training. Classification is done by Softmax once the model is able to distinguish between dominating and low-level features in images. There are various architectures of CNNs like LeNet, AlexNet, VGGNet, GoogLeNet, ResNet, and ZFNet.

Since the 1950s, computer scientists have been trying to make sense of visual data to computers. Kunihiko Fukushima, a Japanese scientist, invented the neocognitron, a very basic image recognition model, using neural networks. A few years later, convolutional neural networks, also called ConvNets, were first introduced in the 1980s by Yann LeCun. LeNet which was named after LeCun could recognize handwritten digits which have an important place in banking to read digits in checks and in postal services to read zip codes. As they could not scale with low-resolution images and less data, they could not perform better in computer vision and AI.

In 2012, AlexNet, an AI model, was developed from the University of Toronto and named after the inventor Alex Krizhevsky who is an expert in image recognition algorithms. The model won the 2012 ImageNet computer vision contest with an 85 percent accuracy which made everyone look into DL again. The ImageNet dataset with complex CNNs is created to perform computer vision tasks that were previously considered as impossible (Cunningham et al. 2008).

9.7.2.2 Unsupervised Learning

Unsupervised learning is used against data that has no historical labels so that the algorithm must figure out on its own to explore and find some pattern in it. Unsupervised learning works well on transactional data, and the popular techniques are self-organizing maps, nearest-neighbor mapping, k-means clustering, and singular value decomposition. Unsupervised learning is used when the information necessary to train is neither classified nor labeled. It understands how the system can infer a function to describe a hidden structure from unlabeled data. The system doesn't figure out the right output, but it explores the data and can draw inferences from datasets to describe hidden structures from unlabeled data. The algorithms take a collection of data that contains only inputs, and identify the structure in the data, like grouping or clustering of data points. The algorithms, therefore, learn from test data that has not been labeled, classified, or categorized. Unsupervised learning algorithms found common features in the data and react based on the presence or absence of similar features in each new item of data.

Cluster analysis is the assignment of a set of observations into subsets (called clusters) that are dissimilar. The clustering techniques make different assumptions on the structure of the data, often defined by similarity metric and evaluated, for example, by internal compactness, or the similarity between members of the same cluster, and separation, the difference between clusters. Other methods are based on estimated density and graph connectivity. They get to know the hidden patterns or data groupings without human intervention which makes it ideal for exploratory data analysis, cross-selling strategies, customer segmentation, image, and pattern recognition. It also uses dimensionality reduction like principal component analysis and singular value decomposition to reduce the number of features. Examples of deep structures that can be trained in an unsupervised manner are neural history compressors and deep belief networks (Hastie et al. 2009).

Autoencoders

Autoencoders are artificial neural networks like unsupervised DL used in dimensionality reduction or as a generative model. They identify patterns from the inputs generating new and provide outputs in the input form. It contains an encoder or recognition network and a decoder or generative network. The input is converted into a simpler, internal representation by an encoder, and the decoder is responsible for producing the output from it.

The input and output layers of autoencoder have the same number of neurons, while the hidden layer has fewer neurons employed to learn the most important features in the data. Due to its lower dimensionality in the hidden layer, the autoencoder is said to be under complete. Autoencoders with multiple hidden layers are called stacked autoencoders. Complex features can be learned with more hidden layers, and sometimes too many hidden layers are likely to overfit the inputs leading to a deficiency in generalization.

Architecture of Autoencoders

The components present in the architecture of autoencoders are:

- **Encoder:** A fully connected feed-forward neural network that compresses the input into a latent space representation and encodes the input image as a compressed representation.
- **Code:** Reduced representation of the input fed into the decoder.
- **Decoder:** Also, a feed-forward network which is responsible for reconstructing the input back to the original dimensions.
  - **Regularized autoencoders:** Prevent from learning the identity function to improve the ability to capture important information.
  - **Sparse autoencoder:** Includes more hidden units to provide sparsity which improves performance and responds to the unique statistical features of the training data.
  - **Denoising autoencoder:** Changes the reconstruction criterion and takes a partially corrupted input which is trained to recover the original undistorted input.
  - **Contractive autoencoder:** Adds a regularizer explicitly with its objective function which forces the model to learn an encoding robust to a slight variation of input values.
  - **Concrete autoencoder:** For discrete feature selection which forces the latent space to consist only of a user-specified number of features.
  - **Variational autoencoder:** Belongs to variational Bayesian methods built with different goals and a completely different mathematical formulation. They are probabilistic autoencoders and are generative autoencoders.

9.7.3 Symbolic AI

Symbolic AI comprises the explicit embedding of human understanding and behavior rules into computer programs. Symbols are type of representations that are used to represent objects. Symbols play an important role in the human cognition and understanding process. The representative objects and abstract concepts are learned and form rules for dealing with them. These rules can be defined in such a way that they capture the knowledge.

Communicating in symbols is one thing that makes us look intelligent and has impacted to play a critical role in the invention of AI. The innovators of AI believed that "every aspect of learning or any other feature of intelligence can in principle be so precisely explained that a machine can be made to simulate it." Therefore, symbolic AI is developed with tools to define and manipulate symbols.

Symbolic AI programs are based on creating structures explicitly with behavior rules, for example, OOP languages that are used to define classes, by specifying properties and organizing to hierarchies. Class instances created perform actions to manipulate properties. Each method performs a series of rule-based commands that might alter the properties of the objects. Symbolic artificial intelligence showed early progress where the rationality of rule-based programs can be easily visualized, communicated, and troubleshoot them. They can be used when the inputs can be directly transformed into symbols, and most computer programs today, including DL applications, follow this.

Efforts have been taken to create complicated symbolic AI systems called expert systems which emphasize the use of hardcoded information and rules to tackle intricate tasks which require an enormous amount of effort by domain experts. Once the problem is generalized, the new rules pop out, and we could not scale this type of AI. Symbolic AI is considered long gone, but in reality, rule-based AI systems are still very essential in today's applications. There have been numerous efforts to combine neural networks and symbolic AI like the Neuro-Symbolic Concept Learner, a hybrid AI system developed by the MIT-IBM Watson AI Lab. It is developed to resolve question-answering problems that can be visualized and can learn new tasks with minimum explainable data and finds it easy to analyze the content of images. Symbolic AI is one of the leading methods for dealing with issues requiring logical reasoning and knowledge representation (Garcez et al. 2012).

9.7.4 Natural Language Processing

Natural language understanding (NLU) recognizes the input in the form of sentences using computer software by enabling human-computer interaction. It allows computers to understand commands without any formalized syntax and respond back to them. The main purpose of NLU is to generate chatbots that can interact with the public without any help. Many major IT companies like Amazon, Apple, Google, and Microsoft, and start-ups run many NLU projects. It analyzes data to understand its meaning by using algorithms to reduce the input into a structured ontology. Intent and entity recognition are the two major concepts of NLU.

Intent recognition is the first and most important part of NLU which identifies the user's sentiment. Entity recognition is a specific type of NLU that focuses on identifying the entities which are named and numeric entities. Named entities are like people, companies, and locations. Numeric entities are numbers, currencies, and percentages. NLP is an emerging technology that derives several forms of AI that are seen in the present times, and its use for designing a seamless, as well as interactive interface among humans and machines, will continue to be the highest primacy for today's and tomorrow's increasingly cognitive applications (Chowdhury 2003).

9.7.5 Reinforcement Learning

Reinforcement learning (RL) is the training to make a sequence of decisions where the agent learns to accomplish a goal in a complex environment. The system faces a situation as a game-like environment and employs trial and error to find a solution for the problem. These results in rewards or penalties from the user for the activities performed with a goal to increase the total reward. The designer fixes the reward policy by rules and the model provides no hints or suggestions on solving. The model identifies how to complete the task and increase the reward, starting from random initialization and completing with sophisticated tactics and superhuman skills. By this technique, reinforcement learning is presently the most successful approach to enhance machine's creativity. Artificial intelligence helps in gathering experience from thousands of parallel gameplays if a reinforcement learning algorithm runs on suitably powerful computer infrastructure.

Weak computer infrastructure limited the scope of reinforcement learning, which was later overcome by Gerald Tesauro's backgammon AI super player created in 1990s. The models which are trained to regulate autonomous cars is a good example of a prospective application of RL. The programmer does not hard-wire anything associated with the task and allows the machine to learn from previous mistakes made by it. In parallel, the only hard-wired element would be the reward function. AlphaGo Zero is the first RL computer program that defeated a world champion in the ancient Chinese game Go and Atari games were developed similar to it. In robotics and industrial automation, RL enables the robot to design an efficient adaptive control system like DeepMind's effort on Deep Reinforcement Learning for Robotic Manipulation with Asynchronous Policy updates. Other applications are engines that summarize text, and speech dialog agents, learning treatment guidelines in healthcare, and online stock trading agents.

The main challenge in RL rests in preparing the simulation environment, which is highly reliant on the task to be performed. The simulation environment is simple in chess, Go, or Atari games, but to drive an autonomous car, the model has to find how to apply the brake or avoid an impact in an environment. This model is then transferred to the training environment to make it ready for the world. Scaling and tweaking the neural is another challenge where the only way to communicate is by rewards and penalties. This can cause the network to forget the learned. Reaching a local optimum by the agents that will improve the reward without performing the designated task is also a challenge.

There is no well-defined division between ML, DL, and reinforcement learning. So, in general, ML is the broadest category, while deep reinforcement learning is the narrowest one and a specialized application of machine and DL techniques that are intended to solve problems in a specific way. The ideas may vary on a slight basis which can amalgamate within projects, as the models are devised to carry out the task in the most effective way possible. So, it is difficult to answer the question "what precisely distinguishes ML, DL, and reinforcement learning?"

Q-learning and state-action-reward-state-action (SARSA) are two frequently used model-free RL algorithms which vary in terms of their exploration strategies. This is a technique where the agent learns the value depending on the action derived from another policy, and SARSA is an on-policy method that learns the value depending on its current action derived from its current policy. These two methods lack generality as they cannot estimate values for unseen states. Deep Q-Networks which use neural networks were developed on behalf of it to estimate Q-values where it can manage discrete, low-dimensional action spaces. Deep Deterministic Policy Gradient is a model-free, off-policy, actor-critic algorithm that can handle high-dimensional, continuous action spaces (Wiering and Van Otterlo 2012).

9.7.6 Cognitive Computing

There are three eras of computing technology, where the first one is in the nineteenth century with the introduction of the programmable computer by Charles Babbage which was used in the navigational calculation and to tabulate polynomial functions. The next era started in 1950 which gave us digital programming computers such as ENIAC an era of modern computing and programmable systems. Cognitive computing represents the third era of computing working basically on DL algorithms and big data analytics to provide insights.

The objective of cognitive computing is to create a model that simulates human processes. The computer can replicate the human brain action which includes self-learning algorithms that use data mining, pattern matching, and natural language processing. Although computers are faster in calculation or processing tasks in comparison to humans, they are still not able to accomplish complete perfectness in tasks such as understanding natural languages or identifying unique objects in an image which are considered as easy for humans.

The cognitive computing process uses a blend of artificial intelligence, neural networks, ML, natural language processing, sentiment analysis, and contextual awareness, to solve everyday problems similar to humans. IBM describes cognitive computing as a sophisticated system that learns at scale, reasoning with purpose, and interacts naturally with humans. According to Gartner, cognitive computing will disrupt the digital sphere as it has the ability to analyze and process large amounts of volumetric data. It helps in employing a computing system that is by far the best in process automation. They are extremely good in gathering, juxtaposing, and cross-referencing information in order to properly assess a situation. In the healthcare industry, cognitive systems such as IBM Watson support physicians in collecting and analyzing data from different sources. They could not replace doctors; it just engages robotic process automation for speeding up data analysis. IBM Watson is also employed in decision-making business processes.

Robotic process automation enhances customer interactions as it can provide contextual information to customers without any interaction and enable customers to engage in business through satisfaction. The basic solution of cognitive function computing plays an important role as a virtual assistant or advisor. Virtual advisor such as Dr. AI by HealthTap is a cognitive solution that depends on individual patients' medical profiles and knowledge collected from 105,000 physicians. Cognitive systems are trying to seep into enterprise systems. In general, the aim of cognitive computing is to develop computing frameworks that can solve complex issues without constant human involvement. (Modha et al. 2011).

9.7.7 Context-Aware Computing

Context-aware computing is an approach of computing in which situational and environmental information about people, places, and things is used to anticipate current requirements and deliver enhanced, situation-aware, and usable content, functions, and experiences. Context-aware computing is a form of computer operation that anticipates situations of usage or, in other words, works in customized ways depending on the context of the user application. This can employ either to a user's activities on the device, or the physical environment in which the device is utilized. In general, engineering concepts are used to satisfy people's expectations in the use of computers or devices. Context-aware computing varies from human-computer interaction by providing greater and more sophisticated functionality that is implemented at runtime, according to input, on the overall context of that specific application.

Context-aware computing includes the design of mobile devices that switch between vertical and landscape orientation, changing screens and illuminates light according to the amount required in the room, and the recent addition of mechanical and sensory elements to reduce damage when they are dropped. Context-aware computing assumes that the computer will need support from users in a particular situation where a person relies on a piece of hardware to perform a task (Schilit et al. 1994).

9.8 Artificial Intelligence in the Healthcare Domain

9.8.1 Introduction

Artificial intelligence has and is proving to be a pioneer in exploring the domain in a whole new aspect, such that advanced studies can be brought up to provide a solution for the existing crisis at hand to the earliest. Though AI was introduced as an infant in 1943 and some claim to be in the 1950s, it was not widely accepted. It took almost 50 years to develop and gain trust for use in the healthcare domain. Its expert system Dendral, the first problem-solving system created for organic chemistry, formed the basis for developing MYCIN. MYCIN is an AI system to detect bacteria in infections such as meningitis or bacteremia and suggests the dosage of antibiotics according to the body weight of the patient. Other systems are like INTERNIST-1, a computer-based diagnostic tool, and casual associational network, a model developed for diagnosis and therapy of glaucoma. Most of the applications in AI deals with data collection, data processing, creating records, NLP and computer vision, robot-assisted surgery, and DL techniques for rare diseases. The main fields at hand are dermatology, radiology, screening, psychiatry, primary care, diagnosis, telemedicine, health records, drug interactions, and in the creation of new ones. The difficulty in improving the technology in this particular domain lies in collecting data and automation. With the improving tech in the field, it had three main reasons to take its place that is to diagnose, to take care of administrative tasks, and to reduce operational costs (Bali et al. 2019).

9.8.2 Market Valuation of AI in the Healthcare Domain

AI's market value varies between a range of four and six billion in the year 2020 by various sources, and its CAGR is expected to be greater than 40% for the upcoming 5 years. The flooding patient-related data, need for personalized medicine, diagnosing diseases at the premature stage, etc. are considered to be the main triggers for AI's peak. With the pandemic effect of 2020, it does not have a fall in the near future (Crawford et al. 2016).

9.8.3 Pioneers in AI-Based Healthcare Domain

9.8.3.1 Microsoft Corporation

Microsoft Corporation is an established American multinational company that has its roots and livelihood in software-related services for computers. Other than that, they have a vast interest in the field of personal computers and computer electronics. It gained its first hold in the field of AI by acquiring Generative Pre-trained Transformer 3, Open AI's natural language model, and integrating it with Microsoft Power apps. It enabled developers to write codes using auto coding feature known as the "low code" feature. Their extension of cloud services to the healthcare industry has proven to connect and manage healthcare data that could engage people in the medical field to provide an efficient service. This service is considered secure and helps a pool of data to knit between them such that we arrive at quicker solutions. The recent update in the cloud platform by including the support of eight additional languages and a virtual patient monitoring system has been a timely help to the users around the ongoing pandemic (Del Sole 2018).

9.8.3.2 Intel

Intel is a technological corporation where it designs, produces, and sells computer products. It deals mostly with hardware technologies for data storage, networking, communications, and computer. It released its first product related to the field of AI by introducing a processor named Nervana NNP-I or Springhill in 2019. This is claimed to work on high workloads with minimal energy. With specific to the field of expertise in hand, Intel had aided in accelerating MRI, genetic research, and radiology to create AI-powered disinfection robots, named Stevie and Violet, and inpatient monitoring technology. Many other contributions are done in analytics, telemedicine, and precision medicine for patients. Their recent interest is in neurology where they have collaborated with Samsung to speed up nerve detection (Alexander et al. 2020).

9.8.3.3 IBM

IBM Corporation produces the whole functioning part of the computer from hardware to software. It crawled into AI by IBM 704 created by Arthur L. Samuel which learned to play checkers from experience. Its notable AI program Watson has till date been the talk of the town for its various features. This cloud-based AI technology was first used for lung cancer patients to provide the correct treatment. Unfortunately, the project dropped due to its failure. Later it had emerged with perfection by improving the technology and had produced various products under six categories of stakeholders which are employer, health plan, government, imaging, life sciences, and provider. For employers, they have a digital health pass, an analytic and data information portfolio, and a mentor with Watson. To create a desirable health plan, they have an analytic and data visualization tool and Watson assistant. For government health and human service analytics, social program management, and Watson care manager imaging can be done by IBM connect Enterprise archive, IBM Connect Access, and Merge PACS. Life sciences can access data from the cloud and use integrated cloud EDC. For providers to offer an insight IBM Micromedex solution, IBM Phytel and healthcare consulting are used (Wiederhold 2017; Tsang et al. 2017).

9.8.3.4 Amazon Web Services

AWS in short is a subsidiary of Amazon that provides cloud computing services and application programming interfaces to its users. Amazon SageMaker is the first product developed where the user can build train and develop an ML model. Without ML experience, the models can be developed and deployed in this platform for their necessity of work. Their other products are advanced text analysis, automatic code reviews, demand forecasting, fraud detection, image, and video analysis, personalized recommendations, etc. Security, compliance, and data privacy solution are provided to the healthcare industry within the scope of HITRUST–CSF (the Health Information Trust Alliance–Common Security Framework). It also provides the infrastructure for businesses to modernize, manages data, and provides analytical solutions by Fast Healthcare Interoperability Resources such as InterSystems IRIS, Amazon HealthLake, and Cerner Health Datalab. Patient experiences AWS by Tele health technology (Yuen et al. 2021).

9.8.3.5 Siemens Healthineers

Siemens Healthineers is a parent company for many medical technology companies which are based out in Germany and one of the top five companies in the world. The company has about 63 AI solutions, over 700 patents in ML, and 275 patents in DL. Being well versed, its theoretical indulgence in the field dates to back the 1980s and 1990s. They couldn't apply the technology processed due to the lack of data and systems that support that large processing. With all the resources available in the modern era, they have emerged as the scholarly organization by bringing out the major possible solutions in AI automation by segregating them into five levels as general assistance, partial automation, conditional automation, high automation, and full automation (Pereira et al. 2020) (Fig. 9.6).

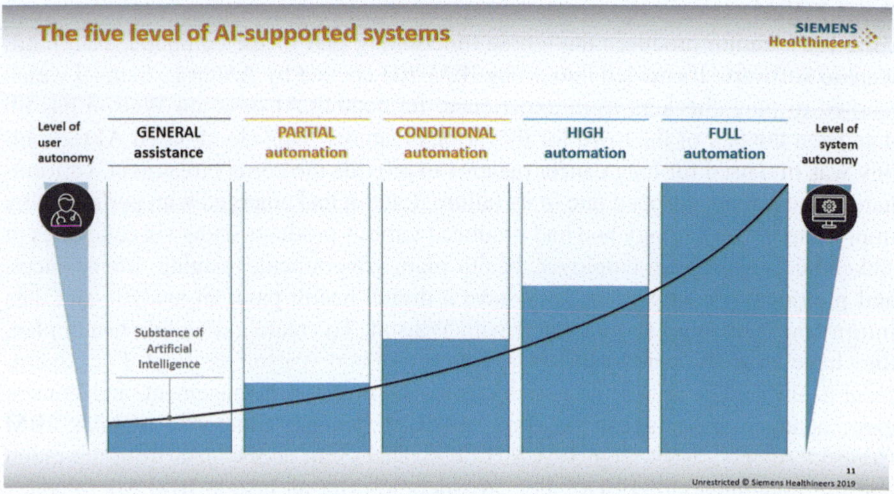

Fig. 9.6 Siemens Healthineers five-level AI-supported systems (Courtesy: Siemens Healthineers)

Their notable contributions are AI-Rad Companion Chest Computed tomography (CT) to reduce the basic cyclic tasks done by the radiologists, pathologists, and clinicians, AI-Pathway Companion to diagnose and provide treatment-related decisions, and the Imaging AI in Practice demonstrated at RSNA. And they are currently working to develop AI solutions for healthcare such that it reaches the full automation level.

9.8.3.6 General Electric

General Electric (GE) has diversified fields of interest like aircraft engines, electrical distribution and motors, energy, finance, healthcare, software, and wind turbines. Their interests lie in hybrid AI, an interdisciplinary technology that can bring together physics and the art of ML. GE healthcare deals with medical imaging, monitoring, biomanufacturing, and cell and gene therapy technologies. These technologies support precision health in diagnostics, therapeutics, and monitoring tools, data analytics, applications, and services. Edison is an intelligent platform created by GE to improve the efficiency of patient outcomes to offer care. Through this critical care suite to analyze images, AIR Recon DL to deliver true fidelity images, Breast Assistant, Embo ASSIST (a 3D virtualization software), SonoCNS for the fetal brain, AIR x to scan a human body, Mural, and much more have been developed (Mastrodicasa et al. 2019).

9.8.3.7 Google

Google has its prime focus on internet-based services and products. The company uses AI in most of the products by mapping one of its functionalities of any form to it. The ML software TensorFlow is the first product developed by the company to enable users to build their models on the open-source library. Google health is enabled with AI to detect disease, data storage, and enabling providers with hardware and insurance. When Google gained Alphabet, three subsidiaries took over the healthcare domain with AI as the main focus. They are Verily, DeepMind, and Calico, which are working to improve their way through a specific set of diseases like eye diseases, diabetes, heart diseases, etc. Their main objective lies in exploring the disease area to provide all over possible solutions through AI, diagnostics, technological tools, and insurance. Using the cancer diagnosis tool that outperformed many radiologists in detecting breast cancer, many such related algorithms are being developed in the same way to implement them as per need (Powles and Hodson 2017).

9.8.3.8 Alibaba

Alibaba specializes in e-commerce, retail, internet, and technology. Its first AI product is a chip to power AI processes called Hanguang 800. About eighty percent of Taobao Mobile users use the AI OS (Artificial Intelligence Online Serving) that runs on AI to integrate recommendations, personalized search, and advertising. They do data processing, data mining and analysis, natural language processing, and a distributed architecture platform. Its contribution to healthcare is by the medical brain developed to apply data intelligence to offer better healthcare services. They take over healthcare management, operational analysis, AI-enabled accessibility, and intelligent scheduling concerns. Its power lies in data integration, data cleaning, and storage, docking, and data platform console construction. It has also released a cancer diagnostic tool called Medical you (Calandra and Favareto 2020).

9.8.3.9 Medtronic

Medtronic is a medical device company that develops and manufactures devices and therapies for the treatment of >30 diseases. The company adapted AI technology to create GI Genius applied in gastrointestinal healthcare for detecting colorectal polyps on the lining of colon or rectum. With the use of AI, data analytics, robot-assisted surgery, and predictive modeling, a pacemaker has been developed that had made the impossible possible with its advanced solution. Sensing technologies are used to power neurostimulators, pacemakers, insulin pumps, and surgical tools. Robotics-assisted surgeries enable doctors to provide patients with standardized care and personalized solution by prediction models. They are still working through their way to provide advanced tech tools in the field of healthcare (Tsang et al. 2017).

9.8.3.10 Baidu's Melody

Baidu specializes in internet-related services, products, and artificial intelligence. Its first AI product is Duer, also known as Chinese apple Siri, which is a voice-activated assistant to provide recommendations to make online purchases. The other products Little fish, Deep Speech, and Deep voice were updates to Duer created to support more of the voice assistant feature. Baidu's Melody is a medical Chabot that helps doctors for diagnosis and its Doctor application released in 2015 (Khan et al. 2018).

9.8.4 Therapeutic Focus Based on AI

The most popular application of AI is ML. According to a survey of Deloitte in 2018, 63% of organizations were already pursuing AI in their businesses. In healthcare, the widely used application of ML is precision medicine. The neural network has been used for applications such as predicting whether a patient would get a specific disease. A common application of DL is the recognition of possible malignant tumors in radiological images. DL is increasingly being used in radiomics and the discovery of clinically important characteristics in imaging data that go beyond what the human eye can visualize. The dominant applications of NLP involve the creation, understanding, and classification of clinical documentation and published research. NLP systems are capable of analyzing unstructured clinical notes on patients, preparing reports, transcribing patient interactions, and conducting conversational AI. Rule-based expert systems have been widely used for clinical decision support purposes over the last couple of decades and are still frequently in use today. Today, many electronic health records providers furnish a set of regulations with their systems. They are also employed in surgery with robots, patient diagnosis, treatment, engagement, and healthcare administrative purposes.

9.8.5 Focus on Cardiovascular Healthcare

9.8.5.1 Introduction on Cardiovascular Diseases

Cardiovascular disease (CVD) is a term to denote the things that affect blood vessels or heart and damages to arteries in the brain, heart, kidneys, and eyes. It generally means the fatty deposits are present in arteries and cause the risk of blood clots. The four major types of CVD are:

- Coronary heart disease (CHD): Affects the arteries that supply blood to the heart muscle.
- Cerebrovascular disease: Affects the arteries that supply blood to the brain leads to Cerebrovascular disease.
- Peripheral arterial disease: Affects the blood arteries that supply the arms and legs.
- Rheumatic heart disease: Streptococcal bacteria causing damages to the heart muscle and valves.

Birth defects that disrupt the normal development and functioning of the heart, caused by deformities of the heart anatomy, are known as congenital heart disease. CVD risk elements include an increase in high blood pressure and cholesterol, diabetes, physical inactivity, smoking, obesity, excessive alcohol intake, and others. The application of AI in cardiovascular disease management has grown exponentially in recent years. Al alone or in combination with laboratory equipment has made footprints in the diagnosis as well as treatment of the patients. AI in combination with laboratory medicine needs to work closely for improving therapeutic outcomes. However, issues like data privacy, patient safety, regulatory bodies, and awareness among the healthcare staff education need to be addressed simultaneously (Yan et al. 2019). The following section will discuss the role of AI in the diagnostic aspects as well as in the treatment process.

9.8.5.2 Statistics About CVD

CVDs are the most frequent cause of death around the world and claimed the lives of 17.9 million individuals worldwide in the year 2019, accounting for 32% of all deaths (http://www.who.int) (Cardiovascular Diseases (CVDs) n.d.). However, heart attacks and strokes are responsible for 85% of these deaths. CVDs are responsible for 38% of the 17 million premature deaths (before the age of 70) caused by noncommunicable diseases in 2019. CVD is one of the major causes of human deaths worldwide and there are various therapies include precision medicine, predictive medicine, preventive medicine and personalized medicine. The following section describes various types of medicine and role of AI in CVD management supporting the healthcare professional leading to better patient life.

9.8.5.3 Precision Medicine

Precision medicine is a personalized medical practice that considers a patient's gene, environment, and lifestyle for prescribing medication. They differ from customized medicine, which divides people into subpopulations affected by a common ailment. They are supplied through the use of multidimensional omics data analysis and systems biology to determine the reason and then provide disease-specific remedies. Precision medicine aids healthcare providers in gaining a better understanding of a variety of factors, including the environment, lifestyle, and heredity. Precision cardiovascular medicine blends unique biological traits, clinical features, biomarkers, human microbiome signatures, and cardiovascular imaging into clinical therapy (Ginsburg and Phillips 2018). Big data analytics catalyze new research and explore towards data-driven knowledge in the future of cardiovascular medicine. In precision medicine, AI helps for the remote follow-ups including prescriptions, counseling, early symptom alerts and helps in the collection of voice data, and the connection of electronic medical record systems with the reduction of clinician labor (Gruson et al. 2020; Krittanawong et al. 2017a). Cognitive computing will support clinicians in making accurate judgments and predicting patient outcomes. Thus, clinicians, in near future, should be familiar with AI technology and practice and in particular for improving cardiovascular disease diagnosis and treatment. With the help of AI, precision medicine will bring customized healthcare management for each individual patient (Ginsburg and Phillips 2018; Ahmed et al. 2020).

The MADIT-CRT (Multicenter Automatic Defibrillator Implantation Trial with Cardiac Resynchronization Therapy) employed an ML algorithm to phenogroup a heart failure cohort based on the complex echocardiographic data and clinical data to predict patients who could benefit from cardiac resynchronization therapy. The study utilized an unsupervised ML algorithm and identified four phenogroups with significant differences in baseline clinical features, biomarker values, and measurements of both ventricular structures (Cikes et al. 2019). Another study used the data of

24,178 patients with acute coronary syndrome (ACS) and ML algorithms for the prediction of major adverse cardiovascular events (MACE) and bleeding. In this study, a super learner ensemble algorithm was applied to select the parameters of 23 different ML algorithms on the basis of 48 selected variables. Among them, the model generated using generalized boosted machine and Elastic Net found a high concordance statistic value of 0.714 for MACE and 0.669 for bleeding, respectively. The outcome of this study also suggests the application of ML in the treatment of antithrombotic medications (Gibson et al. 2020).

The application of AI-based precision medicine has increased largely due to cost reduction in the treatment and an exponential increase in biomedical data. In the future, efforts to create interconnected and curated data sets will improve the accuracy of the prediction using AI algorithms. Many such attempts are already underway, for example, the Precision Medicine Platform was founded by the American Heart Association in collaboration with Amazon Web Services (Kass-Hout et al. 2018). The objective of this platform is to (Dagnino 2021) make data accessible to everyone, (Pannala et al. 2020) easy accessible across orthogonal data sets, (Steinhoff 2021) provide a secure workspace for accessing the power of cloud computing, and (Paschen et al. 2020) a discussion forum to share insights (Krittanawong et al. 2018; Bahado-Singh et al. 2020).

9.8.5.4 Predictive Medicine

Predictive medicine predicts the chance of an individual developing a disease in the future based on the results of specific assessments and analyzing an individual's health history. Blood samples collected from newborn babies for the analysis of possible genetic disorder is a typical example of predictive medicine. Furthermore, risk testing, diagnostic testing, and preconception testing are few examples of classical predictive medicine. Predictive medicine is a well-established practice in the areas of genetic testing, molecular genetic diagnosis, and neurodegenerative disorders. Through the identification of newer signals from a disease, AI can extract novel insights from existing data such as the use of stromal cells for breast cancer prognosis, (Beck et al. 2011) predicting cardiovascular risk factors from retinal fundus photograph, (Poplin et al. 2018), and thickness of retinal layers to assist an earlier diagnosis of dementia (Mutlu et al. 2018). ML algorithms such as Naïve Bayesian (NB) are used in predictive medicine for the diseased condition including asthma, prostate and breast cancer, Alzheimer's disease from genome-wide data, classifier and genetic score for risk prediction, and decision support system for heart disease (Ahmed et al. 2020). Hence the role of AI in predictive medicine is progressing with a tremendous opportunity for the healthcare industry and further improving the overall accuracy of diagnosis, support in the decision for the treatment, and reduces the cost involved (Tuena et al. 2020).

9.8.5.5 Preventive Medicine

Preventive medicine focuses on the health of individuals and communities promoting good health and preventing disease, disability, and death. The experts in the healthcare sector can apply their knowledge and skills in medicine for the betterment of health, improving the quality of life, families, and communities. Preventive medicine aims to avoid disease formation or halting disease progression. A team from MIT has developed a DL algorithm that can quickly identify and screen for early-stage melanoma, a type of malignant tumor (Soenksen et al. 2021). They developed a tool using a deep convolution neural network to identify suspicious pigmented lesions analysis using a dermatological data set that consists of wide-field images (38,283) from 133 patients. They were further validated with a set of wide-field images (135) collected from 68 patients. The model developed using an AI algorithm allows for fast and accurate prediction in pigment lesions. This helps in improving patient treatment plans, the requirement of resources, and prior treatment for melanoma. Researchers are exploring different AI approaches for improving and enhancing care in the treatment of the patient and deliver better results (Rowe and Lester 2020).

9.8.5.6 Personalized Medicine

Personalized medicine refers to medical treatment tailored to a patient's medical history and risk prediction. Although the term personalized medicine is in close association with precision medicine, there is a minor difference. According to the statement by National Research Council, personalized medicine refers to the therapeutic procedure completely specific for an individual patient. It is determined on the basis of patients' particular molecular and genetic profile, which makes them vulnerable to specific diseases. This can also be used to suggest which medical treatments are safe and effective for the patient. However, in the case of precision medicine the treatment can be applicable to group of people. In comparison to a "trial-and-error" approach to disease treatment, physicians can select a therapy with the help of AI that can limit severe side effects and help in reducing the expenses. Personalized medicine has the potential to bring revolution in the identification and treatment of patients in healthcare (Schork 2019). The different algorithms including the NB, ANN, and support vector machine (SVM) are explored for personalized medicine. AI can play an important role in the development of personalized medicines at all relevant phases of the clinical development and implementation of new personalized health products, from finding appropriate intervention targets to testing them for their utility.

A phenotypic personalized medicine (PPM) platform is an alternative to big data, systems biology, and omics-driven personalized medicine. This method identifies the best combination therapy, i.e., the drug and dose, by detecting a relationship between the input stimuli and a phenotypic response based solely on the treated individual's data. Neural networks were employed for correlating multisignal combinations and their effects through a learning process and led to an observation that the relationship between dose and treatment efficacy is represented by a smooth quadratic surface (Al-Shyoukh et al. 2011). PPM aims to maximize the desired outcome via selecting drugs for combination therapy with their initial doses and dynamically recommend the most effective dosing approach with CURATE.AI. CURATE.AI is an AI-derived platform that maps the relationship between an intervention intensity (input) and a phenotypic result (output) for an individual, based exclusively on that individual's data (Blasiak et al. 2020). Drug is the input

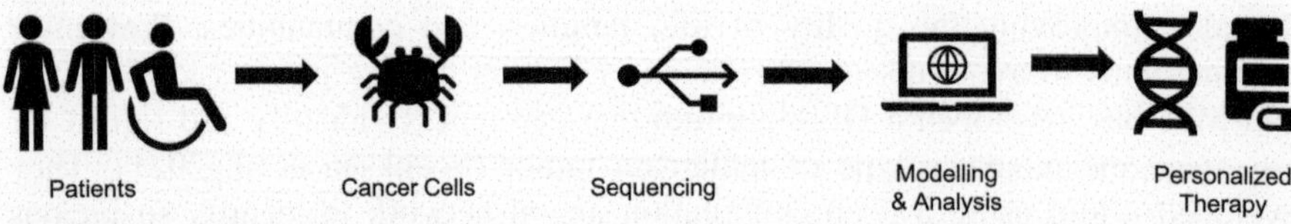

Fig. 9.7 Personalized cancer therapy

with a range of intensities (drug doses) correlated to a measurable output, which is used to calibrate the platform and create a patient's individualized CURATE.AI profile. The creation of an individualized CURATE.AI profile necessitates the collection of a tiny dataset that belongs to an individual and can be used for a variety of indications because of its simplicity and versatility. The CURATE.AI approach's validity has been proved successful for single-drug optimization, combination treatment, oncology, and immunosuppressive, as well as for retrospective and prospective dosing (Lee et al. 2017; Zarrinpar et al. 2016; Pantuck et al. 2018) (Fig. 9.7).

9.8.5.7 Diagnostic Tools

Echocardiography

An echocardiography, commonly known as echocardiogram, cardiac echo, or echo, is a procedure that uses sound waves to create real-time images of the heart. This investigation helps the physician to check the functioning of the heart and valves. There are various techniques of echocardiogram including a transthoracic echocardiogram, contrast echocardiography, fetal echocardiography, transesophageal echocardiography, intracardiac echocardiography, deformation echocardiography (strain rate imaging), stress echocardiography, intravascular ultrasound, and 3D echocardiography.

In echocardiography, models generated using ML approach include features like left ventricular wall thickness and left ventricular ejection fraction for predicting the patient's heart disease condition (Awan et al. 2018). The application of ML is to deliver fully automated support to the healthcare professionals for increasing the accuracy of diagnosis and thereby solving the patient's problem quickly and efficiently. Furthermore, considerable volumes of data are generated in echocardiography such as multidimensional images obtained from 3D echocardiography and speckle tracking, and this could be utilized efficiently with ML approaches (Krittanawong et al. 2017b; Alsharqi et al. 2018) This could help clinicians to make a clear decision for taking care of the patients (Alsharqi et al. 2018; Kusunose 2021; Schuuring et al. 2021; Kusunose et al. 2019).

Electrocardiography

Electrocardiography or electrocardiogram (ECG) is a painless test that examines the electrical activity of the heart. The test provides insight into the physiological and anatomical state of the heart and can also provide important diagnostic signals for systemic diseases. Although the capture of the ECG recording is well standardized and repeatable, human interpretation of the ECG differs substantially depending on experience and competence. CNN has been explored and constructed from large datasets of single-lead ECGs and then applied to the 12-lead ECG in several instances. For example, researchers have applied DNN to generate a model using two million measurements and predicted six different types of abnormalities recorded in 12-lead ECG (Ribeiro et al. 2020). The prediction efficiency of this model with an accuracy (F1-score) >80% and specificity >99% outperforms cardiologists. The results from the study suggest the expansion of these automated interpretations using AI algorithms with different populations will be a very useful diagnostic tool for cardiologists (Hannun et al. 2019).

Cardiac Imaging

Cardiac imaging is a noninvasive imaging technique that uses ultrasound, magnetic resonance imaging, computed tomography, or nuclear medicine imaging to visualize the heart. The imaging techniques are used to identify or observe cardiac disease as well as to assess the anatomy of the heart. Furthermore, coronary catheterization, intravascular ultrasound, and fractional flow reserve techniques are utilized for cardiac imaging. This vast improvement in cardiac imaging techniques resulted in a large amount of imaging data. The available data in combination with AI has the potential to improve the diagnosis of images and patient care (Lim et al. 2020). In particular, the AI algorithms are used to assist in analyzing the images obtained from the techniques and further analysis of structural diseases, such as valvular disease for categorizing and staging (Yan et al. 2019).

The AI algorithms implemented in cardiac imaging are composed of the three basic components, detection, segmentation, and classification (Chen et al. 2020). Guo et al. propose a centerline extraction framework using a DL-based approach multitask fully convolution network (FCN). The approach FCN was used to construct a local centerline distance map of a coronary artery tree object and detecting branch endpoints with a minimal path extractor. The method was validated using a dataset obtained from 620 patients with 400 as the test set. The approach provided an improved success rate of 88.8% compared to 54.3% from the traditional minimal path approach (Guo et al. 2019). In another study, Nasr-Esfahani et al. reported an automatic workflow for segmenting the left ventricle in cardiac MR images which supports physicians to diagnose abnormalities in the heart. The approach first extracts the region of interest, second applies FCN to segment the regions. Finally, the probability output obtained from FCN is applied a roundness function for selecting the proper region. The method was validated with a data set of cardiac MR images from York University with sensitivity value of 87.69 and obtained the best dice score of 87.23 compared with the other methods including active appearance models, active shape model, and with robust estimation method called EM-RASM (Nasr-Esfahani et al. 2018). Similarly, a wide range of AI algorithms are increasingly utilized in healthcare for diagnostic imaging to improve the accuracy in the prediction of cardiovascular risk of patients (Dilsizian and Siegel 2014; Dilsizian and Siegel 2018; Petersen et al. 2019; Gilbert et al. 2020; Al'Aref et al. 2019; Sevakula et al. 2020).

Cardiac Magnetic Resonance Image

Cardiovascular magnetic resonance imaging (CMRI) or cardiac MRI is a noninvasive method of evaluating the function and structure of the cardiovascular system. CMRI is often used in evidence-based diagnostic and treatments and in number of clinical practices and is one of the highly recommended techniques. The application of AI with cardiac MRI brings accurate and easy diagnosis of disease screenings. For example, HeartVista is an AI-driven approach towards MRI and is a one-click autonomous MRI acquisition solution for cardiac exams (HeartVista n.d.; Arafati et al. 2019). The software is developed in association with Siemens and NVIDIA can be used with existing MRI scanners and the AI approach and cloud computing to guide image acquisition. The technology is rapidly offering AI-based automated segmentation solutions and is receiving FDA approval. Another example is Arterys, approved in 2017, to automatically measure and segment ventricles (Arterys n.d.).

Cardiac Electrophysiology

The study of understanding, diagnosing, and treating the electrical processes of the heart is known as cardiac electrophysiology. In cardiac electrophysiology, complex arrhythmias are assessed, symptoms are elucidated, aberrant electrocardiograms are evaluated, the risk of developing arrhythmias in the future is assessed, and the treatment is designed. Atrial fibrillation, bradycardia, tachycardia, ventricular tachycardia, supraventricular tachycardia, ventricular fibrillation, sudden cardiac arrest, long QT syndrome, Wolff-Parkinson-White syndrome, and other arrhythmias can all be detected using the electrocardiogram (ECG or EKG). Blood tests, echocardiogram, stress testing, Holter monitor, event recorder, electrophysiology study, tilt table testing, and implantable loop recorder are all used for diagnosis of disease in a patient.

The increased automated diagnosis of clinical 12-lead ECGs is a significant opportunity for AI in electrophysiology (Feeny et al. 2020). The total therapy time can be reduced to half when the high-risk patients are accurately diagnosed. The current automated ECG diagnosis techniques are ineffective. To enhance time-to-treatment and reduce the workload, DNNs have been used to automate diagnosis or triage ECGs. In large data sets, DNNs can outperform cardiology residents and non-cardiologists and achieve good diagnosis performance. The progress has been achieved in the use of ECG data for predicting atrial fibrillation in sinus rhythm ECGs or screening for hypertrophic cardiomyopathy. Data from intracardiac electrograms before and during defibrillator activation is another fascinating application of DNN algorithms. Analysis of the signals prior to the adverse event may reveal information about the cause of ventricular arrhythmia providing the clinician useful information. Continuous monitoring helps in the detection of asymptomatic cardiac arrhythmias and post-surgical problems. Early detection has the potential to prevent serious adverse effects and provide prompt personalized medicine and healthcare (Muthalaly and Evans 2020).

9.8.5.8 Cardiac Resynchronization Therapy

Cardiac resynchronization therapy (CRT) uses a pacemaker to assist the heartbeat in the appropriate rhythm and timing pattern. The therapy involves the placement of electrodes in the left and right ventricles of the heart, which is controlled by a pacemaker. The timing of the upper heart chambers (atria) and the lower heart chambers (ventricles) is coordinated by the CRT pacemaker. When the heart's two bottom chambers (ventricles) do not beat at the same time, the blood can accumulate in lungs and legs and result in heart failure (Ojo et al. 2017).

In a study, Kalscheur et al. applied six ML algorithms to develop a model which could differentiate mortality or heart failure hospitalization after CRT. The study utilized CRT-pacemaker patients (a total of 481) with 45 features and the generated model validated using CRT-defibrillator patients (a total of 595) found the model generated using random forest predicted patient outcome more precisely (AUC 0.74) in comparison to other models (Kalscheur et al. 2018).

9.8.5.9 Decision Support Tool

A decision support tool, also known as a decision support system (DSS), is a software that supports analysts and authorities for better and faster judgements. It collects and analyzes the massive amounts of data and delivers a comprehensive report for decision-making. AI-based systems can extract insights from many resources and evaluate the data at a deeper level with limited period of time. They can also uncover hidden knowledge in a data by emphasizing facts, persons, entities, and the connection among the data providing a better insight for a decision-making. The five different types of DSS are model-driven, communication-driven, data-driven, document-driven, and knowledge-driven DSS (Power n.d.).

Clinical predictive modeling highly depends on cutting-edge AI approaches to provide physicians with clinical decision support tools (Chen 2020). For predicting clinical deterioration, several machine learning approaches including k-nearest neighbors, decision trees, random forest, SVM, and neural networks outperform the classic approach like logistic regression (Churpek et al. 2016). Clinical DSS's main goal is to identify crucial elements for the categorization or prediction of a problem which are combined to provide an output for helping the physician to clinical conditions and diagnosis. In the field of imaging, the biomedical decision support system helps to select the appropriate test for the patient at an appropriate time (Khan et al. 2018; Massalha et al. 2018; Montani and Striani 2019; Zihni et al. 2020; Setiawan et al. 2020).

9.8.5.10 Models for Various CVD Assessment

Linear regression, logistic regression, decision tree, random forest (RF), support vector machine (SVM), k-nearest neighbors (KNN), and ANN are some examples of supervised learning methods. The most widely utilized algorithms are RF and SVM. There has been a lot of research on applying ML to enhance patient cardiovascular outcomes. With the use of speckle tracking echocardiogram data, Narula and his coworkers constructed an ensemble machine learning model with three classifiers (SVM, RF, and ANN) to automate the differentiation of hypertrophic cardiomyopathy from normal hypertrophy in athletes (Narula et al. 2016). The ML model with speckle tracking echocardiography-based parameters shows that diagnostic ability is comparable to the standard 2D echocardiographic and Doppler-derived parameters often used in clinical practice.

An associative memory classifier (AMC), a cognitive ML classifier, was developed to integrate clinical and imaging data for application in cardiac abnormality identification (Sengupta et al. 2016). The diagnostic AUC for the AMC was improved largely from 0.892 to 0.962 when the four additional echocardiographic variables, viz., end-diastolic septal and posterior wall thickness, Mitral ratio (E/e′), interventricular septum, and posterior wall thickness of the left ventricle, were included to the top 15 speckle tracking echocardiography variables identified from the initial studies. The AMC outperformed other four classifiers, RF, SVM, KNN, and neural networks, selected for comparison.

A DL algorithm, CNN, was used to generate a model on the basis of images and videos from 267 echocardiograms (15 standard views and 12 includes video and still images) stored in UCSF's clinical database. The model classified 15 standard views with an overall accuracy value of 91.7% and 12 video views with better accuracy of 97.8%. The prediction accuracy is significantly higher in comparison to echocardiographers with a value of 79.4% for 15 views obtained from still images (Madani et al. 2018).

Based on the retinal fundus photographs, Poplin and his coworkers developed models using DL for predicting common cardiovascular risk factors (Poplin et al. 2018). The model was trained using retinal fundus photographs from 48, 101 and 236, 234 patients stored in UK Biobank and EyePACS, respectively. The model was validated using the 12,026 (UK Biobank) and 999 (EyePACS) patients' data and predicted major adverse cardiac events with an AUC of 0.70. In another study by Wuang et al., one of the DL approaches, CNN, was applied to detect breast arterial calcifications. The model was developed on the basis of mammograms for automatic identification of breast arterial calcifications and as a coronary artery disease predictor. The model uses a pixel-wise, patch-based technique to distinguish the breast arterial calcification correctly from others and the level of predictions is comparable to human experts (Cho et al. 2021). Table 9.2 lists different machine learning models for the assessment of CVD.

9.8.5.11 Cardiac Transplantation

A heart transplant, also known as a cardiac transplant, is a surgery used to replace a heart in patients who have advanced heart failure or severe coronary artery disease (Deng 2002). The cardiac transplant patients must be monitored intensively for the possible occurrence of acute cellular rejection (ACR). Despite regular intake of immunosuppressants, ACR might occur when recipient T cells recognize donor antigens and destroy cardiac myocytes. Thus, machine learning approaches are applied to differentiate ACR-related myocyte damage from normal situations with reasonable accuracy. The generated model distinguishes myocyte damage from non-myocyte damage with an accuracy value of 94% (Glass et al. 2020).

Table 9.2 Different machine learning models for the assessment of CVD

Model	AUC-ROC	Application	References
Framingham score[a]	0.72	CVD risk score	Alaa et al. (2019)
Cox PH model (7 core variables)	0.73	Risk prediction of CVD	Alaa et al. (2019)
Cox PH model (all variables)	0.75	Risk prediction of CVD	Alaa et al. (2019)
Support vector machines	0.70	Risk prediction of CVD	Alaa et al. (2019)
Neural networks	0.75	Risk prediction of CVD	Alaa et al. (2019)
AdaBoost	0.75	Risk prediction of CVD	Alaa et al. (2019)
Gradient boosting	0.76	Risk prediction of CVD	Alaa et al. (2019)
AutoPrognosis (7 core variables)	0.74	Risk prediction of CVD	Alaa et al. (2019)
AutoPrognosis (369 non-lab. variables)	0.76	Risk prediction of CVD	Alaa et al. (2019)
AutoPrognosis (104 lab. variables)	0.73	Risk prediction of CVD	Alaa et al. (2019)
AutoPrognosis (all variables)	0.77	Risk prediction of CVD	Alaa et al. (2019)
Random forest	0.80	Predict the combined outcome of cardiac arrest	Churpek et al. (2016)
Associative memory classifier (AMC)	0.96	Discrimination of cardiac abnormalities	Sengupta et al. (2016)
Deep learning-based model	0.70	Prediction of major adverse cardiac events	Poplin et al. (2018)
Machine learning based predictive modeling framework	0.84	Prediction of atherosclerosis risk	Okser et al. (2010)
Memetic pattern-based algorithms MPA model	0.87	Prediction of coronary artery disease	Zellweger et al. (2018)
Machine learning based model (LogitBoost)	0.79	Prediction of all-cause mortality	Motwani et al. (2017)

[a] The Framingham score is provided as the reference model for comparative purposes

Myocardial biopsy is the standard procedure for identifying the severity of acute rejection after heart transplantation and the manual elucidation of myocardial biopsy questions the Inter-rater reliability (Cunningham et al. 2006). Thus, the application of AI-based approaches such as CNN which is specialized in feature extraction possibly predicts the inflammation site and the amount of inflammation in the sample. The model trained using pathological images of samples was able to identify inflammatory cells or normal tissue with 91% accuracy (Masuda et al. 2018). This high accuracy prediction suggests AI-based algorithms could support a streamlined procedure in the diagnosis of rejection in the implanted organ (Glass et al. 2020; Goswami 2021).

9.8.5.12 Clinical Predictions/Meta-Analysis

A clinical prediction rule, also called a clinical probability assessment, is an algorithm that utilizes medical signs, symptoms, and other clinical outcomes to predict the likelihood of disease. Recently, a database for clinical prediction models (a total of 796) was published between 1900 and 2012 for predicting CVD (Wessler et al. 2015a). Among them, de novo models (717) are large in number in comparison to recalibration (Hastie et al. 2009) or adaption models (Blasiak et al. 2020) including c-statistic values for 450 de novo models. There are around 800 models for predicting CVD clinically but statistical analysis such as meta-analysis brings together the findings of several scientific investigations when there are more studies addressing the same subject (Wessler et al. 2015b).

For example, atlas-based analysis is used to describe the variation of heart shape in asymptomatic people, quantify preterm ventricular remodeling, and characterize the geometry of the left ventricle after myocardial infarction (Zhang et al. 2014; Cox et al. 2019; Farrar et al. 2016). The analysis helps to understand the shape and motion differences of the heart of disease patients from normal people (Medrano-Gracia et al. 2013; Gilbert et al. 2018).

Interestingly with the support of AI, it is possible to predict an individual will die due to a heart disorder. A study by a research group from UK's Medical Research Council developed a tool by utilizing a data set of MRI scans of 256 patients and their blood test reports. Additionally, they collected 30,000 data points representing the direction and magnitude of systolic excursion for the endocardial vertex in the right ventricle. With this information as well as the health records of the patients, AI learned the abnormalities and calculated the life of patients. According to the risk levels the tool has good prediction accuracy for the patients who would still be alive for the next 5 years and in particular after 1-year survival is 80% in comparison to the doctors with a value of 60% (Dawes et al. 2017). In another study, Motwani et al. applied ML approaches to predict the probability of mortality of patients with suspected coronary artery disease. The ML approach includes automatic feature selection and model building using boosted ensemble algorithm; the model validated with ten-fold cross-validation shows better results for ML than current clinical or coronary computed tomography angiography metrics (Motwani et al. 2017).

A meta-analysis of the 344 studies reporting ML models generated to predict CVD found good accuracy with AUC values between 0.8 and 0.9 (Krittanawong et al. 2020). Similarly values are shown for coronary artery disease and stroke prediction compared to heart failure or cardiac arrhythmias. Among different algorithms compared in this study, boosting algorithms and SVM found promising results in the prediction of coronary artery disease (CAD) and stroke risk, and in particular, SVM outperforms boosting in image recognition problems and vice versa in omics-related problems (Sun et al. 2013; Ogutu et al. 2011).

9.8.5.13 Robots Used in CVS/Chatbots

As a part of a digital revolution, in the late 1990s, robots are incorporated into CVD. The application of robotic-assisted devices in CVD provides an opportunity for healthcare professionals to provide a precise and less invasive diagnosis

(Torregrossa et al. 2021). Among different robotic devices, HeartLander is a commonly used one and helps the doctors in heart surgeries without opening the patient's chest (Rasmussen et al. 2021). It can easily detect, map, and treat issues on the surface of the heart. The other common surgeries where robots are involved are repairing the mitral valve, atrial septal defects, bypass surgeries of coronary arteries, and myxoma resection. Vascular surgeons utilize robot-assisted surgeries for the treatment of aortic, visceral, and peripheral artery diseases and electrophysiology of atrial fibrillation ablation (Bonatti et al. 2014).

In percutaneous coronary intervention (PCI), novel robotic-assisted and the remote-controlled procedure is a potential solution to the healthcare professionals where the radiation exposure and orthopedic injuries can be reduced (Stevenson et al. 2021; Chakravartti and Rao 2019). A robot-assisted PCI system allows a protected control console to manipulate coronary guide wires, balloons, and stents during PCI. Corindus Vascular Robotics now offers CorPath 200 and CorPath GRX two robotic systems. The CorPath 200 has been thoroughly tested in clinical trials and applied as robotic-assisted PCI.

Another robotic-assisted system is the da Vinci surgical system applied in CVD for performing minimal invasive surgery. Compared to traditional endoscopic surgery, the da Vinci system allows complex surgical movements (Pugin et al. 2011). The Sensei and Magellan robotic navigation system from Hansen Medical was developed for remote, catheter control in electrophysiological procedures (Peters et al. 2018). The specific algorithms of an AI implemented in these robotic systems will significantly improve safety, efficacy, quality, and efficiency in the surgeries and treatment of CVD (Lanfranco et al. 2004; De Silva et al. 2020).

9.8.5.14 Interventional Procedure Assistance

An interventional procedure is a minimal invasive diagnostic, therapeutic, or biopsy procedure. It is a non-surgical treatment generally applied in CVD to open narrowed coronary arteries to increase the blood supply to the heart. When a blockage is identified, the interventional procedure is performed during a diagnostic cardiac catheterization or it can be done after the confirmation of CAD and the recovery time from catheterization is much shorter than that of surgery.

AI has a significant impact on automated clinical decision-making, medical imaging analysis, and interventional procedures and has the capability of influencing the practice of interventional cardiology (Sardar et al. 2019). The AI improves real-time clinical decision-making, streamlines and standardizes catheter-based procedures through advanced robotics. In interventional cardiology, the automated analysis of clinical data utilizing AI-based algorithms will provide patient-specific risk scores for procedural planning. The models generated using AI will predict the risk of death, bleeding, contrast nephropathy, and stroke (Krittanawong et al. 2017a). Furthermore, the application of cognitive computing can predict the clinical outcome and prioritizes the tasks in the catheterization laboratory. The integration of periprocedural data helps to monitor recovery, complications, and advise the required period for therapy (Hashimoto et al. 2018).

9.8.5.15 Approved Medical Devices Developed Using AI

The FDA has authorized 64 AI-based medical devices and algorithms (Fig. 9.8) (Benjamens et al. 2020). The first FDA approval came in 2016, followed by three more before the end of 2017, and later in the period 2018–2020, thirteen, ten, and two devices were approved. Among the approval devices, radiology and cardiology are the two main medical specializations and other devices are focusing on Internal medicine/endocrinology, neurology, ophthalmology, emergency medicine, and oncology. In the field of cardiology, there has been significant progress with the approval of four medical devices and algorithms.

Another recent study shows that since 2015 there has been a significant increase in approval for AI-based medical devices. The database of medical devices applying AI is updated including European Medicines Agency (EMA) and US FDA. A total of 240 and 222 devices have been approved in Europe and the USA, respectively (Muehlematter et al. 2021). Majority of the approved devices focus on radiology

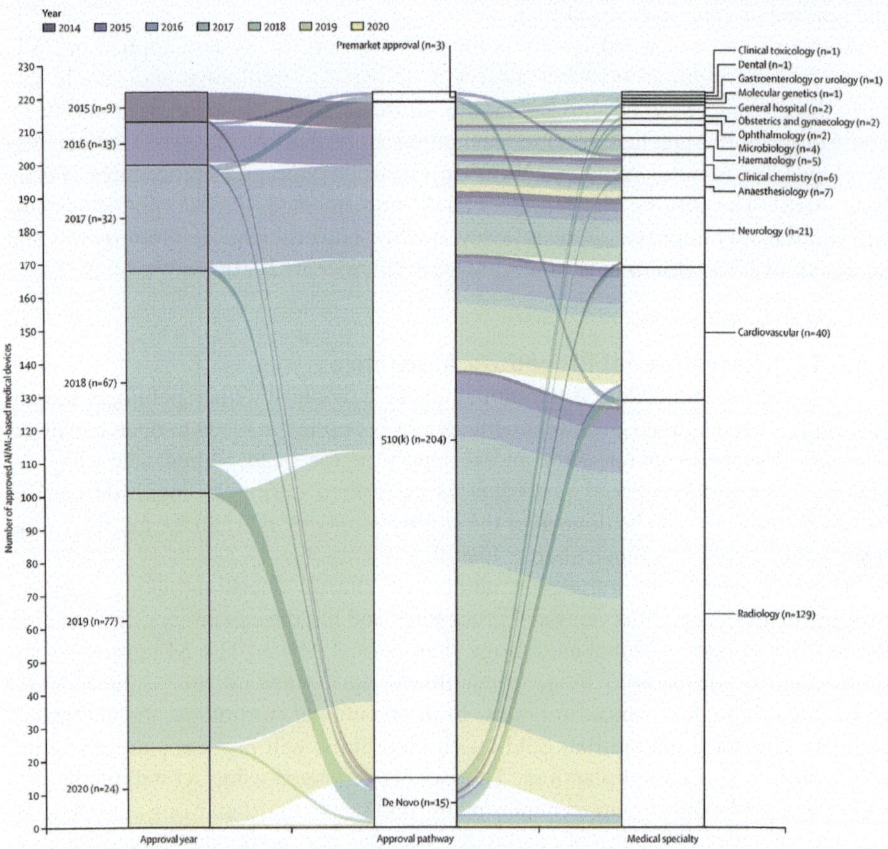

Fig. 9.8 FDA approvals for AI-based devices in medicine (Muehlematter et al. 2021)

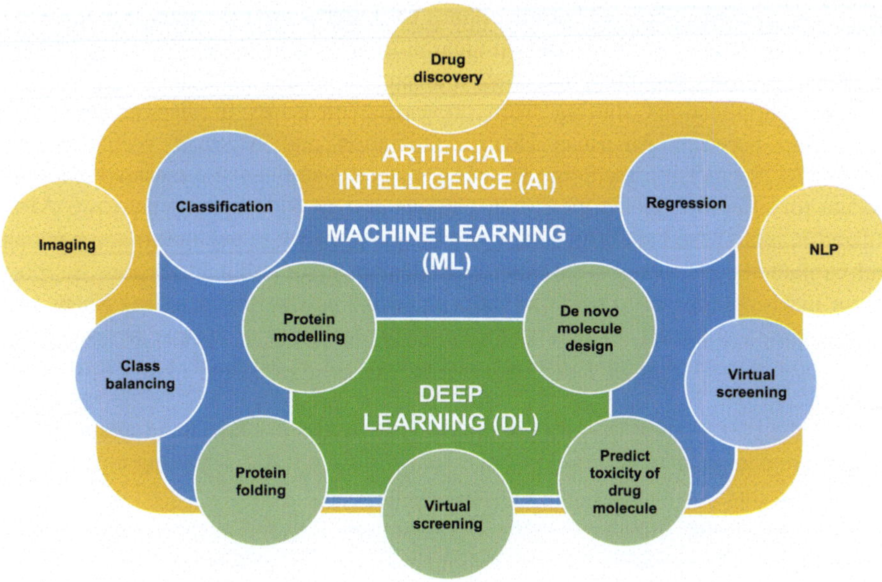

Fig. 9.9 AI in drug discovery

(129 devices) and followed by cardiology (40 devices) highlighting the significance of AI in healthcare management.

9.8.6 Drug Discovery and Development Using AI

AI can be utilized in drug discovery, including drug design, chemical synthesis, drug screening, and drug repurposing (Fig. 9.9). AI can help in the identification of hits and lead molecules for the optimization of drug structure. The regularly employed ML couldn't handle data of millions of compounds. A quantitative structure-activity relationship model can predict thousands of physicochemical parameters but predictions of complex biological properties, such as the efficacy and adverse effects of compounds, are not possible. Recently developed DL and relevant modeling studies in AI help in the safety and efficacy evaluations of drug molecules from big data analysis (Paul et al. 2021).

9.9 Conclusion and Future Perspective

In this chapter, the significance of AI, the typology, and the evolution of AI has been introduced and its application in healthcare are discussed with a special focus on cardiovascular management. The ML and DL techniques are explained in detail with supervised and unsupervised learning along with their subtypes ANN, RNN, CNN, and Autoencoders as well as symbolic learning, natural language processing, reinforcement learning, cognitive and context-aware computing. AI can be employed in healthcare applications starting from preventive measures to disease cure. The partnership between American Heart Association and Amazon Web Services brought AI in precision medicine allowing to access and analyze massive amounts of data for effective cooperation with patients and provide better treatment. AI is utilized in atlas-based analysis to describe shape variance in asymptomatic persons and explain the patterns of remodeling following myocardial infarction. Support vector machine appears to perform better in stroke patients by handling linear and nonlinear data efficiently. The advancement in these kinds of approaches and their application in cardiovascular disease management from diagnosis to treatment are highlighted.

With the vast amount of information, AI is expected to understand analysis and solve more complicated issues closer to actual clinical questions leading to the better decision for disease management. Researchers have started venturing AI in the entire field for achieving higher success. The da Vinci robotic system for coronary revascularization was recently authorized by FDA. The application of AI in the diagnosis sector such as in echocardiography, electrocardiography, cardiac magnetic resonance image, cardiac electrophysiology, and in decision support tool increases high efficiency. The cardiology field has shown significant progress, with four AI-based medical devices and algorithms approved by FDA.

AI being an emerging and fast-growing technology also needs to address issues from regulations such as the safety and efficacy of the technology. Recently, FDA has provided guidance for assessing AI systems. In the future, AI will find its primary application in the healthcare sector with a significant role for supporting medical professionals. AI might manage patients individually depending upon their condition and if required then a specialist physician can be approached. In the current pandemic situation due to SARS-CoV-2, AI can perform its task by interacting with humans remotely and solve their problems. In the future healthcare, technologies need to provide permanent solutions to save lives and reduce risk on patients due to human errors. AI has an important role in healthcare and cardiovascular diseases solving complex problems of individuals by constant training with the data. AI contributes significantly in healthcare with an objective to improve the quality of human life and minimize errors with the support and expertise of healthcare professionals.

References

Ahmed Z, Mohamed K, Zeeshan S et al (2020, 2020) Artificial intelligence with multi-functional machine learning platform development for better healthcare and precision medicine. Database: baaa010

Alaa AM, Bolton T, Di Angelantonio E et al (2019) Cardiovascular disease risk prediction using automated machine learning: a prospective study of 423,604 UK Biobank participants. PLoS One 14:e0213653

Al'Aref SJ, Anchouche K, Singh G et al (2019) Clinical applications of machine learning in cardiovascular disease and its relevance to cardiac imaging. Eur Heart J 40:1975–1986
Alexander A, Jiang A, Ferreira C et al (2020) An intelligent future for medical imaging: a market outlook on artificial intelligence for medical imaging. J Am Coll Cardiol 17:165–170
Alsharqi M, Woodward WJ, Mumith JA et al (2018) Artificial intelligence and echocardiography. Echo Res Pract 5:R115–Rr25
Al-Shyoukh I, Yu F, Feng J et al (2011) Systematic quantitative characterization of cellular responses induced by multiple signals. BMC Syst Biol 5:88
Arafati A, Hu P, Finn JP et al (2019) Artificial intelligence in pediatric and adult congenital cardiac MRI: an unmet clinical need. Cardiovasc Diagn Ther 9:S310–Ss25
Arterys (n.d.) Arterys receives FDA clearance for the first zero-footprint medical imaging analytics cloud software with deep learning for cardiac MRI. https://www.prnewswire.com/news-releases/arterys-receives-fda-clearance-for-the-first-zero-footprint-medical-imaging-analytics-cloud-software-with-deep-learning-for-cardiac-mri-300387880.html. Accessed 15 July 2021
Awan SE, Sohel F, Sanfilippo FM et al (2018) Machine learning in heart failure: ready for prime time. Curr Opin Cardiol 33:190–195
Bahado-Singh RO, Vishweswaraiah S, Aydas B et al (2020) Precision cardiovascular medicine: artificial intelligence and epigenetics for the pathogenesis and prediction of coarctation in neonates. J Matern Fetal Neonatal Med 4: 1–8
Bali J, Garg R, Bali RT (2019) Artificial intelligence (AI) in healthcare and biomedical research: why a strong computational/AI bioethics framework is required? Indian J Ophthalmol 67:3
Barto AG, Sutton RS. Reinforcement learning in artificial intelligence. Adv Psychol. 121: Elsevier; 1997. p. 358–386
Beck AH, Sangoi AR, Leung S et al (2011) Systematic analysis of breast cancer morphology uncovers stromal features associated with survival. Sci Transl Med:3:108ra13
Benjamens S, Dhunnoo P, Meskó B (2020) The state of artificial intelligence-based FDA-approved medical devices and algorithms: an online database. NPJ Digit Med 3:118
Blasiak A, Khong J, Kee T (2020) CURATE.AI: optimizing personalized medicine with artificial intelligence. SLAS Technol 25:95–105
Bonatti J, Vetrovec G, Riga C et al (2014) Robotic technology in cardiovascular medicine. Nat Rev Cardiol 11:266–275
Calandra D, Favareto M (2020) Artificial intelligence to fight COVID-19 outbreak impact: an overview. Eur J Soc Impact Circ Econ 1:84–104
Cardiovascular Diseases (CVDs) (n.d.) World Health Organization (WHO). https://www.who.int/news-room/fact-sheets/detail/cardiovascular-diseases-(cvds). Accessed 15 July 2021
Chakravartti J, Rao SV (2019) Robotic assisted percutaneous coronary intervention: hype or hope? J Am Heart Assoc 8:e012743
Chen L (2020) Overview of clinical prediction models. Ann Transl Med 8:71
Chen C, Qin C, Qiu H et al (2020) Deep learning for cardiac image segmentation: a review. Front Cardiovasc Med 7:25
Cho SY, Kim SH, Kang SH et al (2021) Pre-existing and machine learning-based models for cardiovascular risk prediction. Sci Rep 11:8886
Chowdhury GG (2003) Natural language processing. Annu Rev Inf Sci Technol 37:51–89
Churpek MM, Yuen TC, Winslow C et al (2016) Multicenter comparison of machine learning methods and conventional regression for predicting clinical deterioration on the wards. Crit Care Med 44:368–374
Cikes M, Sanchez-Martinez S, Claggett B et al (2019) Machine learning-based phenogrouping in heart failure to identify responders to cardiac resynchronization therapy. Eur J Heart Fail 21:74–85
Cox DJ, Bai W, Price AN et al (2019) Ventricular remodeling in preterm infants: computational cardiac magnetic resonance atlasing shows significant early remodeling of the left ventricle. Pediatr Res 85:807–815

Crawford K, Whittaker M, Elish MC et al (2016) The AI now report. The Social and Economic Implications of Artificial Intelligence Technologies in the Near-Term
Cunningham KS, Veinot JP, Butany J (2006) An approach to endomyocardial biopsy interpretation. J Clin Pathol 59:121–129
Cunningham P, Cord M, Delany SJ (2008) Supervised learning. In: Cord M, Cunningham P (eds) Machine learning techniques for multimedia: case studies on organization and retrieval. Berlin, Heidelberg, Springer, pp 21–49
Dagnino A (2021) Data analytics in the era of the industrial internet of things. Springer, Cham
Dawes TJW, Marvao A, Shi W et al (2017) Machine learning of three-dimensional right ventricular motion enables outcome prediction in pulmonary hypertension: a cardiac MR imaging study. Radiology 283:381–390
De Silva K, Myat A, Strange J et al (2020) Iterative improvement and marginal gains in coronary revascularisation: is robot-assisted percutaneous coronary intervention the new hope? Interv Cardiol 15:e18
Del Sole A (2018) Introducing Microsoft cognitive services. Microsoft computer vision APIs distilled. Springer, Apress, Berkeley, CA, pp 1–4
Deng MC (2002) Cardiac transplantation. Heart 87:177–184
Dilsizian SE, Siegel EL (2014) Artificial intelligence in medicine and cardiac imaging: harnessing big data and advanced computing to provide personalized medical diagnosis and treatment. Curr Cardiol Rep 16:441
Dilsizian ME, Siegel EL (2018) Machine meets biology: a primer on artificial intelligence in cardiology and cardiac imaging. Curr Cardiol Rep 20:139
El-Amir H, Hamdy M (2020) Deep learning pipeline. Springer, Apress, Berkeley, CA
Esteva A, Robicquet A, Ramsundar B et al (2019) A guide to deep learning in healthcare. Nat Med 25:24–29
Farrar G, Suinesiaputra A, Gilbert K et al (2016) Atlas-based ventricular shape analysis for understanding congenital heart disease. Prog Pediatr Cardiol 43:61–69
Feeny AK, Chung MK, Madabhushi A et al (2020) Artificial intelligence and machine learning in arrhythmias and cardiac electrophysiology. Circ Arrhythm Electrophysiol 13:e007952
Garcez ASA, Broda KB, Gabbay DM (2012) Neural-symbolic learning systems: foundations and applications. Springer, London
Gibson WJ, Nafee T, Travis R et al (2020) Machine learning versus traditional risk stratification methods in acute coronary syndrome: a pooled randomized clinical trial analysis. J Thromb Thrombolysis 49:1–9
Gilbert K, Forsch N, Hegde S et al (2018) Atlas-based computational analysis of heart shape and function in congenital heart disease. J Cardiovasc Transl Res 11:123–132
Gilbert K, Mauger C, Young AA et al (2020) Artificial intelligence in cardiac imaging with statistical atlases of cardiac anatomy. Front Cardiovasc Med 7:102
Ginsburg GS, Phillips KA (2018) Precision medicine: from science to value. Health Aff (Millwood) 37:694–701
Girasa R (2020) AI as a disruptive technology. In: Artificial intelligence as a disruptive technology: economic transformation and government regulation. Palgrave Macmillan, Cham, pp 3–21
Glass C, Davis R, Xiong B et al (2020) The use of artificial intelligence (ai) machine learning to determine myocyte damage in cardiac transplant acute cellular rejection. J Heart Lung Transplant 39:S59
Goswami R (2021) The current state of artificial intelligence in cardiac transplantation. Curr Opin Organ Transplant 26:296–301
Gruson D, Bernardini S, Dabla PK et al (2020) Collaborative AI and laboratory medicine integration in precision cardiovascular medicine. Clin Chim Acta 509:67–71
Guo Z, Bai J, Lu Y et al (2019) DeepCenterline: a multi-task fully convolutional network for centerline extraction. CoRR abs/1903.10481
Haleem A, Javaid M, Khan IH (2019) Current status and applications of artificial intelligence (AI) in medical field: an overview. Curr Med Res Pract 9:231–237

Hamet P, Tremblay J (2017) Artificial intelligence in medicine. Metabolism 69s:S36–s40

Hannun AY, Rajpurkar P, Haghpanahi M et al (2019) Cardiologist-level arrhythmia detection and classification in ambulatory electrocardiograms using a deep neural network. Nat Med 25:65–69

Hashimoto DA, Rosman G, Rus D et al (2018) Artificial intelligence in surgery: promises and perils. Ann Surg 268:70–76

Hastie T, Tibshirani R, Friedman J (2009) Unsupervised learning. The elements of statistical learning: data mining, inference, and prediction. Springer, New York, NY, pp 485–585

HeartVista (n.d.) HeartVista announces the first AI-driven, One Click™ Autonomous MRI Solution to Enable Faster and More Efficient Scans that Benefit Patients, Technicians, and Radiologists. https://www.prnewswire.com/news-releases/heartvista-announces-the-first-ai-driven-one-click%2D%2Dautonomous-mri-solution-to-enable-faster-and-more-efficient-scans-that-benefit-patients-technicians-and-radiologists-300754565.html. Accessed 15 July 2021

History of Artificial Intelligence (n.d.) The University of Queensland, Queensland, https://qbi.uq.edu.au/brain/intelligent-machines/history-artificial-intelligence. Accessed 15 July 2021

Husbands P, Harvey I, Cliff D et al (1997) Artificial evolution: a new path for artificial intelligence? Brain Cogn 34:130–159

Jiang F, Jiang Y, Zhi H et al (2017) Artificial intelligence in healthcare: past, present and future. Stroke Vasc Neurol 2:230–243

Jordan MI, Mitchell TM (2015) Machine learning: trends, perspectives, and prospects. Science 349: 255–260

Kalscheur MM, Kipp RT, Tattersall MC et al (2018) Machine learning algorithm predicts cardiac resynchronization therapy outcomes: lessons from the COMPANION trial. Circ Arrhythm Electrophysiol 11:e005499

Kass-Hout TA, Stevens LM, Hall JL (2018) American Heart Association precision medicine platform. Circulation 137:647–649

Khan RS, Zardar AA, Bhatti Z (2018) Artificial intelligence based smart doctor using decision tree algorithm. arXiv preprint arXiv:180801884

Krittanawong C, Zhang H, Wang Z et al (2017a) Artificial intelligence in precision cardiovascular medicine. J Am Coll Cardiol 69:2657–2664

Krittanawong C, Tunhasiriwet A, Zhang H et al (2017b) Deep learning with unsupervised feature in echocardiographic imaging. J Am Coll Cardiol 69:2100–2101

Krittanawong C, Johnson KW, Hershman SG et al (2018) Big data, artificial intelligence, and cardiovascular precision medicine. Expert Rev Precis Med Drug Dev 3:305–317

Krittanawong C, Virk HUH, Bangalore S et al (2020) Machine learning prediction in cardiovascular diseases: a meta-analysis. Sci Rep 10:16057

Kusunose K (2021) Steps to use artificial intelligence in echocardiography. J Echocardiogr 19:21–27

Kusunose K, Haga A, Abe T et al (2019) Utilization of artificial intelligence in echocardiography. Circ J 83:1623–1629

Lanfranco AR, Castellanos AE, Desai JP et al (2004) Robotic surgery: a current perspective. Ann Surg 239:14–21

Lee DK, Chang VY, Kee T et al (2017) Optimizing combination therapy for acute lymphoblastic leukemia using a phenotypic personalized medicine digital health platform: retrospective optimization individualizes patient regimens to maximize efficacy and safety. SLAS Technol 22:276–288

Lim LJ, Tison GH, Delling FN (2020) Artificial intelligence in cardiovascular imaging. Methodist Debakey Cardiovasc J 16:138–145

Madani A, Arnaout R, Mofrad M et al (2018) Fast and accurate view classification of echocardiograms using deep learning. NPJ Digit Med 1

Massalha S, Clarkin O, Thornhill R et al (2018) Decision support tools, systems, and artificial intelligence in cardiac imaging. Can J Cardiol 34:827–838

Mastrodicasa D, Albrecht MH, Schoepf UJ et al (2019) Artificial intelligence machine learning-based coronary CT fractional flow reserve (CT-FFRML): impact of iterative and filtered back projection reconstruction techniques. J Cardiovasc Comput Tomogr 13:331–335

Masuda H, Miyagawa S, Minakami F et al (2018) Abstract 10989: artificial intelligence utilizing pathological diagnosis of acute rejection after cardiac transplantation. Circulation 138: A10989-A

McNally L, Brown SP, Jackson AL (2012) Cooperation and the evolution of intelligence. Proc Biol Sci 279:3027–3034

Medrano-Gracia P, Cowan BR, Bluemke DA et al (2013) Atlas-based analysis of cardiac shape and function: correction of regional shape bias due to imaging protocol for population studies. J Cardiovasc Magn Reson 15:80

Modha DS, Ananthanarayanan R, Esser SK et al (2011) Cognitive computing. Commun ACM 54: 62–71

Montani S, Striani M (2019) Artificial intelligence in clinical decision support: a focused literature survey. Yearb Med Inform 28:120–127

Motwani M, Dey D, Berman DS et al (2017) Machine learning for prediction of all-cause mortality in patients with suspected coronary artery disease: a 5-year multicentre prospective registry analysis. Eur Heart J 38:500–507

Muehlematter UJ, Daniore P, Vokinger KN (2021) Approval of artificial intelligence and machine learning-based medical devices in the USA and Europe (2015-20): a comparative analysis. Lancet Digit Health 3:e195–e203

Muthalaly RG, Evans RM (2020) Applications of machine learning in cardiac electrophysiology. Arrhythm Electrophysiol Rev 9:71–77

Mutlu U, Colijn JM, Ikram MA et al (2018) Association of retinal neurodegeneration on optical coherence tomography with dementia: a population-based study. JAMA Neurol 75:1256–1263

Narula S, Shameer K, Salem Omar AM et al (2016) Machine-learning algorithms to automate morphological and functional assessments in 2D echocardiography. J Am Coll Cardiol 68: 2287–2295

Nasr-Esfahani M, Mohrekesh M, Akbari M et al, (2018) Left Ventricle Segmentation in Cardiac MR Images Using Fully Convolutional Network. 2018 40th Annual International Conference of the IEEE Engineering in Medicine and Biology Society (EMBC); 18–21 July 2018

Nilsson NJ, Nilsson NJ (1998) Artificial intelligence: a new synthesis. Morgan Kaufmann, San Francisco, CA

Ogutu JO, Piepho HP, Schulz-Streeck T (2011) A comparison of random forests, boosting and support vector machines for genomic selection. BMC Proc 5(Suppl 3):S11

Ojo A, Tariq S, Harikrishnan P et al (2017) Cardiac resynchronization therapy for heart failure. Interv Cardiol Clin 6:417–426

Okser S, Lehtimäki T, Elo LL et al (2010) Genetic variants and their interactions in the prediction of increased pre-clinical carotid atherosclerosis: the cardiovascular risk in young Finns study. PLoS Genet 6:e1001146

Pannala R, Krishnan K, Melson J et al (2020) Emerging role of artificial intelligence in GI endoscopy. Gastrointest Endosc 92:1151–1152

Pantuck AJ, Lee D-K, Kee T et al (2018) Modulating BET bromodomain inhibitor ZEN-3694 and enzalutamide combination dosing in a metastatic prostate cancer patient using CURATE.AI, an artificial intelligence platform. Adv Therap 1:1800104

Paschen U, Pitt C, Kietzmann J (2020) Artificial intelligence: building blocks and an innovation typology. Bus Horiz 63:147–155

Paul D, Sanap G, Shenoy S et al (2021) Artificial intelligence in drug discovery and development. Drug Discov Today 26:80–93

Pereira VM, Cancelliere NM, Nicholson P et al (2020) First-in-human, robotic-assisted neuroendovascular intervention. J Neurointerv Surg 12:338–340

Peters BS, Armijo PR, Krause C et al (2018) Review of emerging surgical robotic technology. Surg Endosc 32:1636–1655

Petersen SE, Abdulkareem M, Leiner T (2019) Artificial intelligence will transform cardiac imaging-opportunities and challenges. Front Cardiovasc Med 6:133

Poplin R, Varadarajan AV, Blumer K et al (2018) Prediction of cardiovascular risk factors from retinal fundus photographs via deep learning. Nat Biomed Eng 2:158–164

Power D (n.d.) Types of decision support systems (DSS), The Global Development Research Center. https://www.gdrc.org/decision/dss-types.html. Accessed 15 July 2021

Powles J, Hodson H (2017) Google DeepMind and healthcare in an age of algorithms. Heal Technol 7:351–367

Pugin F, Bucher P, Morel P (2011) History of robotic surgery: from AESOP® and ZEUS® to da Vinci®. J Visc Surg 148:e3–e8

Ramesh A, Kambhampati C, Monson JR et al (2004) Artificial intelligence in medicine. Ann R Coll Surg Engl 86:334

Rasmussen ET, Shiao EC, Zourelias L et al (2021) Coronary vessel detection methods for organ-mounted robots. Int J Med Robot 17(5):e2297

Ribeiro AH, Ribeiro MH, Paixão GMM et al (2020) Automatic diagnosis of the 12-lead ECG using a deep neural network. Nat Commun 11:1760

Rowe JP, Lester JC (2020) Artificial intelligence for personalized preventive adolescent healthcare. J Adolesc Health 67:S52–Ss8

Sardar P, Abbott JD, Kundu A et al (2019) Impact of artificial intelligence on interventional cardiology: from decision-making aid to advanced interventional procedure assistance. JACC Cardiovasc Interv 12:1293–1303

Schilit B, Adams N, Want R, (1994) Context-aware computing applications. 1994 First Workshop on Mobile Computing Systems and Applications; IEEE

Schork NJ (2019) Artificial intelligence and personalized medicine. Cancer Treat Res 178:265–283

Schuuring MJ, Išgum I, Cosyns B et al (2021) Routine echocardiography and artificial intelligence solutions. Front Cardiovasc Med 8:648877

Sengupta PP, Huang YM, Bansal M et al (2016) Cognitive machine-learning algorithm for cardiac imaging: a pilot study for differentiating constrictive pericarditis from restrictive cardiomyopathy. Circ Cardiovasc Imaging 9:e004330

Setiawan NA, Venkatachalam PA, Hani AFM (2020) Diagnosis of coronary artery disease using artificial intelligence based decision support system. arXiv preprint arXiv:200702854

Sevakula RK, Au-Yeung WM, Singh JP et al (2020) State-of-the-art machine learning techniques aiming to improve patient outcomes pertaining to the cardiovascular system. J Am Heart Assoc 9:e013924

Soenksen LR, Kassis T, Conover ST et al (2021) Using deep learning for dermatologist-level detection of suspicious pigmented skin lesions from wide-field images. Sci Transl Med 13

Steinhoff J (2021) Machine learning and fixed capital: the contemporary AI industry. Automation and autonomy: labour, capital and machines in the artificial intelligence industry. Springer, Cham, pp 133–170

Stevenson A, Kirresh A, Ahmad M et al (2021) Robotic-assisted PCI: the future of coronary intervention? Cardiovasc Revasc Med

Strong A (2016) Applications of artificial intelligence & associated technologies. Science [ETEBMS-2016] 5

Sun T, Wang J, Li X et al (2013) Comparative evaluation of support vector machines for computer aided diagnosis of lung cancer in CT based on a multi-dimensional data set. Comput Methods Prog Biomed 111:519–524

Torregrossa G, Amabile A, Oosterlinck W et al (2021) The epicenter of change: robotic cardiac surgery as a career choice. J Cardiac Surg 36(10):3497–3500

Tsang L, Kracov DA, Mulryne J et al (2017) The impact of artificial intelligence on medical innovation in the European Union and United States. Intellect Prop Technol Law J 29:3–11

Tuena C, Semonella M, Fernández-Álvarez J et al (2020) Predictive precision medicine: towards the computational challenge. In: Pravettoni G, Triberti S (eds) P5 eHealth: an agenda for the health technologies of the future. Springer, Cham, pp 71–86

Wessler BS, Lai YHL, Kramer W et al (2015a) Clinical prediction models for cardiovascular disease. Circ Cardiovasc Qual Outcomes 8:368–375

Wessler BS, Lai YL, Kramer W et al (2015b) Clinical prediction models for cardiovascular disease: tufts predictive analytics and comparative effectiveness clinical prediction model database. Circ Cardiovasc Qual Outcomes 8:368–375

Wiederhold BK (2017) What can behavioral healthcare learn from digital medicine? Mary Ann Liebert, Inc., New Rochelle, NY

Wiering MA, Van Otterlo M (2012) Reinforcement learnin

Yan Y, Zhang JW, Zang GY et al (2019) The primary use of artificial intelligence in cardiovascular diseases: what kind of potential role does artificial intelligence play in future medicine? J Geriatr Cardiol 16:585–591

Yu KH, Beam AL, Kohane IS (2018) Artificial intelligence in healthcare. Nat Biomed Eng 2:719–731

Yuen M-C, Chu SY, Wong CH et al (2021) Development and Pilot Test for Stuttering Self-Monitoring Solution using Telehealth. 2021 International Conference on COMmunication Systems & NETworkS (COMSNETS); IEEE

Zarrinpar A, Lee DK, Silva A et al (2016) Individualizing liver transplant immunosuppression using a phenotypic personalized medicine platform. Sci Transl Med 8:333ra49

Zellweger MJ, Tsirkin A, Vasilchenko V et al (2018) A new non-invasive diagnostic tool in coronary artery disease: artificial intelligence as an essential element of predictive, preventive, and personalized medicine. EPMA J 9:235–247

Zhang X, Cowan BR, Bluemke DA et al (2014) Atlas-based quantification of cardiac remodeling due to myocardial infarction. PLoS One 9:e110243

Zihni E, Madai VI, Livne M et al (2020) Opening the black box of artificial intelligence for clinical decision support: a study predicting stroke outcome. PLoS One 15:e0231166

The manufacturer's authorised representative in the EU is Springer Nature Customer Service Centre GmbH, Europaplatz 3, 69115 Heidelberg, Germany. If you have any concerns regarding our products, please contact ProductSafety@springernature.com

Printed and bound by CPI Group (UK) Ltd, Croydon, CR0 4YY
30/09/2026
02225248-0001